Learning: A Very Short Introduction

VERY SHORT INTRODUCTIONS are for anyone wanting a stimulating
and accessible way into a new subject. They are written by experts, and
have been translated into more than 45 different languages.

The series began in 1995, and now covers a wide variety of topics in
every discipline. The VSI library now contains over 500 volumes—a Very
Short Introduction to everything from Psychology and Philosophy of
Science to American History and Relativity—and continues to grow in every
subject area.

Titles in the series include the following:

Mark Haselgrove

LEARNING

A Very Short Introduction

OXFORD
UNIVERSITY PRESS

OXFORD
UNIVERSITY PRESS

Great Clarendon Street, Oxford, OX2 6DP,
United Kingdom

Oxford University Press is a department of the University of Oxford.
It furthers the University's objective of excellence in research, scholarship,
and education by publishing worldwide. Oxford is a registered trade mark of
Oxford University Press in the UK and in certain other countries

© Mark Haselgrove 2016

The moral rights of the author have been asserted

First edition published in 2016

Impression: 2

Published in the United States of America by Oxford University Press
198 Madison Avenue, New York, NY 10016, United States of America

British Library Cataloguing in Publication Data
Data available

Library of Congress Control Number: 2016931581

ISBN 978-0-19-968836-4

Printed in Great Britain by
Ashford Colour Press Ltd, Gosport, Hampshire

Contents

List of illustrations

Learning

Chapter 1
What is learning?

Learning is a phenomenon that can be observed across the animal kingdom. Complex creatures such as humans can do it, as can insects such as honey bees and ants. If you read this book, remember some of the information that is contained within it, and then maybe go on to use it, you will have demonstrated an ability to learn. If you have a pet cat or dog, then you will have probably noticed that they suddenly become very attentive towards you once you begin making them a meal. I am reasonably confident that my cat, for example, was not born with a tendency to follow me around the kitchen like a shadow just before the time that she normally has her breakfast. I therefore suspect that she has learned this behaviour. Learning is not only restricted to your waking hours either, you are even learning about the outside world when you are asleep. The neurobiologist Arnat Arzi and her colleagues showed that sleeping people could learn to change the amount that they sniffed during a particular sound once an experimenter had established it as a signal for a nasty smell.

Yet, what is learning? Despite the many examples of learning that readily come to mind, it turns out to be relatively difficult to pin down an entirely satisfactory definition of the phenomenon. One intuitive definition of learning might be to suggest that it is related, in some way, to *the acquisition of information*. However, a moment's reflection reveals that this definition is unsuitable: a library acquires information, yet it would be difficult to argue that it has therefore learned. Perhaps then we should qualify this statement a bit and say that learning is *the acquisition of information by a biological organism*. This definition is better, but it is too restrictive: it excludes the learning that occurs in machines or computers, which we experience in everyday life (think of the ability of the search engine Google or your smartphone to predict what you are going to type). It is probably fair to say that there is no generally accepted definition of learning. However, many of the crucial aspects of learning can be understood as *a relatively permanent change in behaviour as a consequence of experience*. This definition emphasizes the relationship between experience (such as experience based on the environment) and a change in behaviour. However, by defining learning in terms of a change in behaviour we are still being restrictive. For example, this book may be read, and all of the information within in it remembered, but the reader may decide to never do anything about it. Does this mean that the information that has been acquired has not been learned? From the perspective of the scientific study of learning, we potentially might have to say *yes*, because no matter how convinced you are that you have acquired the information in this chapter just by reading it, you are making this judgement by relying on introspection—an examination of one's own thoughts or feelings. From a scientific perspective, this has been argued to be a dangerous thing to do as introspection can be a poor method for establishing psychological principles. It is far better, instead, to rely on measuring something from an organism—and that is why incorporating behaviour into a definition of learning is important. Responses can be measured and linked to differences in environmental experience and, so long as we have designed our experiments properly, we can then reach some kind of conclusion about whether learning has occurred and, more interestingly, discover the principles that allowed the learning to take place.

Why do organisms learn?

Evolution has provided animals with reflexes that are present from birth and which are expressed, on many occasions, without the necessity of learning. Reflexes can also be automatic, and fast. For example, in a reflex arc, a sensory neuron connects to the spine and from there onto a motor neuron without having to pass through the brain. This permits sensory perception (such as the sensation of burning in the hand) to be quickly acted upon (withdrawing the hand from a fire) to reduce tissue damage without the potentially limiting influence of the brain—which may be rather busy processing other information.

Reflexes, then, clearly, have their benefits. However, they also have their limitations, because most organisms inhabit environments that have predictive relationships between events. For example, fire is typically bright, has a smoky smell about it, and gets increasingly warm the closer you get to it. By learning to link these features of fire with a burn, animals can simply learn to keep away from things that are associated with fire—thus they avoid being burnt rather than having to reflexively withdraw from fire every time it is encountered. However, the predictive relationships between events in the world are not always permanent: environments may change with time. For example, the cues for finding nourishment in the summer may be rather different to the cues that are needed to find food during the winter. Thus, to avoid starving between the seasons, animals must be able to modify their behaviour as a consequence of their experiences with, in this case, the environment. It may be recalled that this was the very definition of learning that was provided just a moment ago.

The purpose of this book is to provide *a very short introduction* to the properties and principles of a number of different types of learning across a variety of different species. We will begin by looking at one of the simplest forms of learning, habituation, before moving on to two forms of learning that, for about a century, have dominated the study of learning in animals—classical and instrumental conditioning. We will then move on to associative learning, which is one of the dominant explanations for how learning takes place, and how it can be applied, as well as investigating how learning can sometimes be counterproductive, before finally addressing the question of whether learning is in fact as simple as some psychologists have made out—or whether a higher order, more *cognitive*, explanation of learning is required. Throughout this discussion, examples of learning from both human and non-human animals will be described. This will help to clarify some of the psychological concepts that will be introduced, as well as to demonstrate how our understanding of learning in animals has been applied to learning in humans.

Habituation

Habituation is, perhaps, the simplest form of learning that one can conceive of, and refers to a reduction in the vigour (or the likelihood) of behaviour as a consequence of the simple repetition of a stimulus. For example, in the corridor outside my office is an air conditioning unit. When this turns on, making a low humming sound, my attention is initially captured by this stimulus and I make what is known as an orienting response towards it: I look at it. But as the air conditioning unit continues to hum, I become less likely to orient towards it. Indeed, after a while, I hardly even notice it. I have become habituated to the sound of the air conditioning unit. Habituation takes place not only to responses that are elicited by sounds, and it is not only confined to humans; the amount of time that a rat spends orienting towards an illuminated light decreases when the light is repeatedly turned on. Interestingly, even single-celled organisms demonstrate habituation. In an experiment reported in 1906 by the zoologist Herbert Spencer Jennings, it was shown that paramecia will contract when they are touched. However, after repeatedly being touched, the number of touches that were required to make the paramecia contract increased to between twenty and thirty.

Despite seeming to be a relatively simple effect, habituation can also provide animals with a way in which they can engage in apparently complex behaviour. For example, habituation can explain why a preference to music might change over time. This is a behaviour that is quite typical of humans but which—perhaps surprisingly—has also been demonstrated in the humble laboratory rat. Henry Cross and his colleagues showed in the late 1960s that rats, when given a choice, avoided music by Schoenberg relative to music by Mozart. Following this initial test, the rats were exposed to music by Schoenberg for twelve hours a day, for fifty-two days—a high level of exposure by any standards. Following this exposure, the rats were once again given an opportunity to determine whether they would hear music by Mozart or Schoenberg. In this second test, a small preference for the music by Schoenberg over Mozart was evident. With experience, it seems that rats can come to prefer Schoenberg. Despite this behaviour appearing complex, it can be conceived of as a demonstration of habituation. The first test revealed that the rats avoided music by Schoenberg when given a choice; and this avoidance response could be based upon a very simple feature of the music (e.g. a preponderance of a particular note that rats find unpleasant). With repeated exposure to the music, this response diminished so that, during the second test, the avoidance response to music by Schoenberg was at a similar level to that for music by Mozart.

Sadly, most studies of habituation in animals do not use classical music as stimuli. Instead, many studies have focused on the habituation of rats' neophobic (a fear or dislike of anything new) responses to novel flavours—that is to say their unwillingness to consume foods that they haven't encountered before. For example, Michael Domjan conducted an experiment and showed that, even if they were thirsty, rats, upon encountering a flavour for the first time, would drink only very little of the liquid (about 5 millilitres (ml)—even if its flavour had been sweetened. After repeated exposure to the flavour over ten days, however, this avoidance response reduced with habituation, and the rats' consumption of the flavour increased to around 15–20 ml. Habituation in cases of neophobia is a good demonstration of how the interaction between learning and innate behaviour can be particularly functional. The cautious reaction to novel foods displayed by rats at first limits the possibility of being poisoned by unfamiliar items—which could be potentially deadly. Habituation, however, allows the animal to learn to widen its diet to include other foods.

Although habituation is a widespread effect, and clearly constitutes an example of learning, a form of learning that has received significant attention from psychologists is conditioning, of which there are two kinds: classical conditioning and instrumental conditioning. We will look at these two forms of learning in turn.

Classical conditioning

Many people have heard of classical conditioning through one route or another. If asked what it is, they will probably describe an experiment performed many years ago by Ivan Pavlov, in which dogs salivated to the sound of a bell. For the most part, this description is entirely correct; however, it is typically about as far as most people's knowledge about conditioning goes. Less well known is that Pavlov's experiment was just one of many conducted by him and his team of scientists, and that many more conditioning experiments have been carried out by other psychologists and neuroscientists since Pavlov's time, many of whom have studied conditioning using different procedures and animals in order to gain a better understanding of how learning works. It is probably not surprising, however, that most of these researchers have not followed Pavlov's lead and studied drooling dogs. That said, many psychologists continue to use experimental designs that were first developed by Pavlov as well as a scientific terminology that Pavlov developed. In order to gain a better understanding of what classical conditioning is, I will introduce four key terms that Pavlov himself used. These four terms allow psychologists to describe the stimuli and responses of *any* classical conditioning experiment. At first, it might seem needlessly confusing to use abstract terms instead of the actual stimuli and responses from a conditioning experiment. However, as we will see now, and later on in this book, it has the advantage of allowing psychologists to identify the key features of a conditioning experiment, and compare or contrast them with different studies whilst using a set of common phrases.

In order to describe these four terms, it is convenient to refer to Pavlov's original conditioning experiment with dogs (see Figure 1), the most important feature of which was probably the *unconditioned stimulus* (which can be abbreviated to *US*). This is a stimulus that has intrinsic biological significance to the animal, and in Pavlov's experiments was often food. This stimulus is called the unconditioned stimulus because it *unconditionally* elicits a response: that is to say, it can elicit a response in the animal without the need for any training—it is an instinctual response. The response that the unconditioned stimulus elicits is called the *unconditioned response* (which can be abbreviated to *UR*), and in Pavlov's experiments this was salivation—eating food made his dogs drool. In his experiments, Pavlov sounded a bell just before

Pavlov's phrase	Definition
Unconditioned stimulus (US)	A biologically significant event (e.g. food or pain)
Unconditioned response (UR)	The response to the US
Conditioned stimulus (CS)	A previously neutral stimulus (e.g. a bell, or light) that acquires a response by being paired with the US
Conditioned response (CR)	The response to the CS

1. **Pavlov's confusing (but ultimately useful) terminology.**

giving food to the dog on a number of occasions (which psychologists often refer to as 'trials'). At first, the sound of the bell evoked little responding in the dog—little more than, say, a turn of its head. Pavlov's key discovery, however, was that after the ringing of the bell had been paired with providing the food a number of times, the sound of the bell itself made the dogs salivate—even when food itself was not being given to the dog. Pavlov called the response that was acquired by the ringing bell a *conditioned response* (which can be abbreviated as *CR*) because its existence was conditional upon the pairing of the sound of the bell with the unconditioned stimulus (in this case, food). Finally, Pavlov termed the sound of the bell ringing the *conditioned stimulus* (which can be abbreviated as *CS*) in order to identify it as a stimulus that could only elicit a response because of its pairing with the unconditioned stimulus.

The way in which Pavlov classified stimuli and responses indicates that classical conditioning is an example of learning—at least in the way it was defined earlier in this book: the behaviour of Pavlov's dogs changed in a relatively permanent fashion (i.e. a conditioned response was acquired) as a consequence of experience (the pairing of the conditioned stimulus with the unconditioned stimulus).

Two examples of classical conditioning

Since Pavlov's discovery of the conditioned response, many thousands of experiments have demonstrated classical conditioning in a whole variety of species: from simple organisms such as the sea slug, *Aplysia californica*, to more complex species such as humans (see Figure 2(a) and (b)). Now, neither of the two experiments that will be described measured drops of saliva from the mouths of slugs or humans while a bell was being rung; entirely different measures of conditioned responding were taken to entirely different stimuli. However, because Pavlov provided us with a terminology that helps us to compare across very different

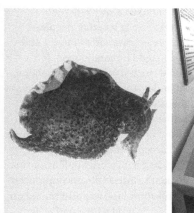

2. (a) A sea slug; and (b) some humans. They appear very different, but both are susceptible to classical conditioning.

experiments, we can identify the key components of these experiments and understand their design.

In the first experiment, Thomas Carew, Robert Hawkins, and Eric Kandel, a Nobel laureate, conducted a study with sea slugs—marine organisms that have particularly large nerve cells, making them ideal candidates for the study of the neural basis of learning (see Figure 2(a)). In one experiment, the researchers paired a slight touch (the conditioned stimulus) of the sea slug's siphon, which is a tube-like structure, with a mild electric shock (the unconditioned stimulus) that was delivered half a second later to the slug's tail. Before the first conditioning trial, the light touch to the slug's siphon had little effect—the siphon withdrew into the slug's body for only about ten seconds. However, after a number of trials in which the light touch was paired with the shock, the light touch came to elicit a withdrawal response that lasted about fifty seconds, even though the tail shock was not presented. The withdrawal response that was elicited by touching the slug's siphon is the conditioned response.

9

Irene Martin and Archie Levey conducted a very different experiment. They were interested in seeing whether people's blinking—a response that helps spread tears across the eye and remove irritants from the cornea—could be conditioned. If it could be conditioned then it would suggest an adaptive function, in which the blink response can anticipate the arrival of an irritant and protect the eye from it rather than simply respond to the irritant. Martin and Levey illuminated coloured lights that were mounted on a panel placed in front of adult humans. One of these lights was turned on for 800 milliseconds (the conditioned stimulus), the end of which coinciding with a brief puff of air that was directed towards the participant's cornea (the unconditioned stimulus). Another light—a control stimulus—was also turned on for 800 milliseconds, but it was never paired with a puff of air. Martin and Levey observed that at the start of the experiment neither of the two lights resulted in the participant blinking. However, after the lights had each been presented twenty times participants did begin to blink prior to the delivery of the puff of air—but only during the presentation of the light that predicted it—the control light did not elicit a blink. This blink to the light that preceded the puff of air was the conditioned response. It seems that blinking can be conditioned—which is quite a useful function.

The experiments conducted by Carew and colleagues and by Martin and Levey are very different to each other in many ways—they measure different responses to different stimuli in very different animals. However, they do have a significant feature in common—in both experiments the unconditioned stimulus was an unpleasant event—specifically, a shock to the sea slug's tail in Carew's experiment, and a puff of air to the eye in Martin and Levey's experiment with humans. In contrast, Pavlov's experiment used an unconditioned stimulus that was a pleasant event—food. We can use this difference between the types of unconditioned stimulus as a way in which to distinguish between different types of classical conditioning experiments. On the one hand we have *aversive conditioning*, using an unpleasant event, as we have already described, and on the other hand we have *appetitive conditioning*, in which the motivational significance of the unconditioned stimulus is positive. We have already encountered an example of appetitive conditioning in Pavlov's experiment in which food was the unconditioned stimulus with dogs, and appetitive conditioning is also commonly demonstrated in a variety of other mammals and birds using food as the unconditioned stimulus. However, appetitive conditioning is not only limited to experiments in which food is the unconditioned stimulus; not surprisingly, water can also be used as an unconditioned stimulus to elicit a conditioned response of licking in thirsty rats. Sex can also serve as an unconditioned stimulus (although in this case the conditioned response measured was neither drooling nor licking). In an experiment conducted by Michael Domjan and his colleagues, male Japanese quail were exposed to a red light before they were allowed to have sex with a female quail. After several pairings of the red light (the conditioned stimulus) with copulation (the unconditioned stimulus) the quail began to spend more time near the red light during its illumination (a conditioned response). They also approached and mounted the female quicker than a control group (which had been exposed to the red light and female quail, but never together). In later experiments, Domjan and his team also showed that males released more sperm during copulation after being exposed to a conditioned stimulus for sex. This result reveals, very directly, the potential adaptive significance of learning.

Learned flavour aversions

The classical conditioning experiments that we have discussed so far have all arranged for multiple pairings of the conditioned and unconditioned stimuli; and for each of these pairings, the two stimuli were presented relatively closely together in time. However, a moment's reflection reveals that learning can take place about the relationship between two stimuli even when they are separated by quite a long interval. For example, for most of my life, I have been unable to drink the relatively sweet, dark-coloured beer called brown ale: either the smell or a small sip of it turns my stomach, and I immediately put some distance between me and the offending beverage. The source of my aversion to brown ale was a night's overindulgence during a field trip to the Lake District. After a number of hours of drinking the ale I felt very nauseous, and suffered a bad hangover the next day.

In all likelihood you too will have acquired an aversion to some beverage in your time—either as a consequence of genuine illness or, like me, overindulgence. At first blush the acquisition of flavour aversion learning seems to bear all the hallmarks of classical conditioning: I drank something (the conditioned stimulus), was later ill (the unconditioned stimulus), and the subsequent aversion to the drink is a conditioned response. The crucial part of the preceding sentence, however, is 'later'. Several hours intervened between my drinking the brown ale and my subsequent illness. Ordinarily, an interval this long would be sufficient to prevent conditioning from taking place between, say, a light and a puff of air to the eye, but not, it seems, between a flavour and illness.

An experiment conducted by James Smith and David Roll demonstrated clearly the tolerance that flavour aversion learning has to the passage of time. Rats in an experimental group were given the opportunity to drink a sucrose solution before then being placed in a chamber and exposed to a dose of X radiation that was sufficiently strong to make them feel nauseous. Control rats received exactly the same treatment but the X radiation was not delivered. In both the experimental and control groups, different rats had different intervals between drinking sucrose and being placed in the chamber. Figure 3 shows the results of the final test stage of Smith and Roll's experiment in which the rats were given a choice between water and sucrose solution the next

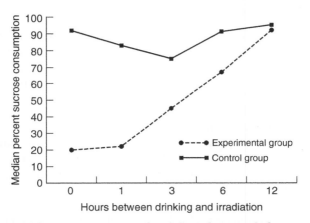

3. **Learned taste aversion. Even after six hours has passed after drinking sucrose solution, rats will still associate this flavour with an illness.**

day. During the test, the rats that had been placed immediately into the chamber and irradiated after drinking sucrose (zero hours on the horizontal axis of the graph) showed a significant avoidance of sucrose relative to the control group. However, rats also continued to avoid the sucrose (relative to the control group) when an interval of up to six hours was interposed between drinking and irradiation. Only when half a day had passed between drinking and the induction of nausea did the data from the experimental group come to resemble the control group.

There are two features of this experiment that are notable. First, as we have observed already, the learned aversion that was acquired by the rats in the experimental groups tolerated a substantial delay between consumption of the conditioned stimulus (sucrose) and the unconditioned stimulus (X radiation). Second, the aversion to the sucrose was acquired after only a single pairing of the drink with illness. A number of psychologists have suggested that these two features of flavour aversion tell us that it stands out as a special form of conditioning—which, ordinarily, is intolerant to intervals between the conditioned and unconditioned stimuli, and which can typically require a number of pairings of these two stimuli before a conditioned response is reliably observed. There is good reason for thinking this way too—in order to avoid poisoning it makes evolutionary sense for an animal to be able to acquire an aversion to a potentially dangerous food quickly, as well as in the case of a delay between eating something toxic and feeling the effects of it. However, other psychologists have countered this claim. They have argued that although flavour aversion learning is tolerant of long delays between the conditioned and unconditioned stimulus, just like other forms of conditioning, it is still the case that increasing the interval between the food and the illness disrupts learning—thus it is still sensitive to the passage of time, just on a different scale. Furthermore, other forms of aversive conditioning can also result in the expression of a conditioned response after only one trial—such as fear conditioning in rats. And, finally, flavour aversion learning is also susceptible to the kinds of stimulus competition effects that will be described in Chapter 3 (e.g. blocking). Together, these observations provide scant support for the idea that flavour aversion learning is, in some way, special.

Losing a conditioned response

One should not be under the impression that once a conditioned response has been acquired, it remains with the individual person or animal for life. Conditioned responses can also be removed, and one of the most widespread ways of doing this is to repeatedly present the conditioned stimulus without the unconditioned stimulus after conditioned responding has been established. What happens, under these circumstances, is that responding weakens—an effect called *extinction*. Consider a classical conditioning experiment with rats conducted in my laboratory. Nine times a day for twelve days, a tone was turned on for ten seconds. As it was turned off a small food pellet was delivered to

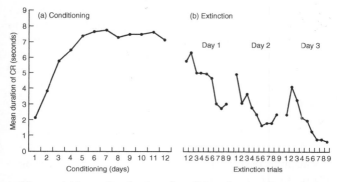

4. The acquisition and extinction of conditioned responding to a tone that was paired with food in laboratory rats.

the rats. As you can see in Figure 4(a), over time, the rats spent more and more time investigating the part of the conditioning apparatus where food would soon be delivered during the tone—the acquisition of the conditioned response. Once this training was complete we continued to present the tone but omitted the delivery of food on each trial for three days. The result, shown in Figure 4(b), is that this conditioned response progressively weakened—extinction. Interestingly, between each day of extinction conditioned responding recovered slightly, despite the fact that no more conditioning had been given in between these days. This effect is called spontaneous recovery, and will be returned to in Chapter 3. Extinction is a well established property of learning, both in classical conditioning and in instrumental conditioning (to which we will turn our attention shortly). It is also a tremendously useful property of learning, for it forms the basis of a psychotherapy called 'exposure therapy', which can be used to remove unwanted behaviours that have been acquired through experience, such as fears and phobias and, as we shall see later, drug craving.

Classical conditioning will be returned to in subsequent chapters, where we will look at what is learned during the procedure, the sorts of variables that seem to affect it, and how it might underpin some examples of unwanted behaviour in people. It is time now to move to the second form of conditioning that is frequently studied by psychologists, and which has a subtle but important difference to classical conditioning. In classical conditioning, the unconditioned stimulus is delivered to the animal irrespective of what the animal does. Thus, it didn't matter whether Pavlov's dogs drooled or not, food was delivered to the dogs by the experimenter after the conditioned stimulus. A rather different procedure for studying conditioning, however, makes the delivery of food (for example) dependent on the behaviour of the animal. This type of learning is referred to as *instrumental conditioning*.

Learning to control the world—instrumental conditioning

Edward Thorndike was the first person to systematically investigate how an animal's behaviour might change as a consequence of its behaviour being related to environmental events. Where Pavlov studied dogs, Thorndike studied cats (although he conducted some experiments with dogs as well). He placed cats into a box (see Figure 5), outside of which was a bowl of food. In order to escape from the box and eat the food, the cat had to perform a response—such as depressing a switch that opened the door of the box. At first, the cat would take quite a time before it made the appropriate response—maybe scratching at the floor or grooming itself instead; once it did make the correct response, however, it was able to escape from the box and was rewarded with a few moments' access to the food before being returned to the box. What Thorndike observed was a reduction in the time it took the cat to escape from the box. At first, the cats may have taken between 60 and 120 seconds to escape, but after about a dozen or so trials, they were escaping in around ten to twenty seconds.

Because the time taken by the cats to escape from the box reduced gradually rather than abruptly, Thorndike argued that the

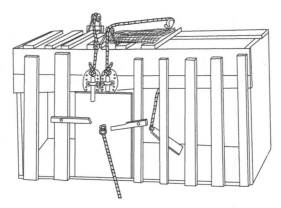

5. Thorndike's puzzle box, which was used to investigate instrumental conditioning in cats.

behaviour reflected a gradual, incremental process of learning rather than, for example, a moment of insight. In fact, Thorndike proposed a specific law of conditioning, *the law of effect*:

> Of several responses made to the same situation, those which are accompanied or closely followed by satisfaction to the animal will, other things being equal, be more firmly connected with the situation. (Thorndike, *Animal Intelligence*, p. 244)

What Thorndike is proposing here is that in his experiment with cats, the food served to strengthen a connection (or association) between (a) the stimuli that the animal could perceive in its environment, and (b) the response that it makes. Thus, the learning that takes place is between a stimulus and a response (known as 'S-R learning'). According to Thorndike the food only served to bolster the connection between a stimulus and a response, and he called events that increase the probability of a response when presented after it 'reinforcers'. We shall return to the question of *what* is learned during learning in Chapter 2—and examine whether Thorndike's analysis of instrumental conditioning is accurate. Other animals also demonstrate instrumental conditioning, and a vast array of animals including rats, pigeons, fish, and various invertebrates can be trained to press levers or pull at chains, peck upon keys, swim through hoops, or dig in sand when such responses are followed by reinforcement. Even humans will acquire a response if it is followed by a reward, as anybody who is paid to go to work will know.

One of the more common forms of instrumental conditioning used in the animal learning laboratory is the acquisition of a lever press response. Figure 6 shows a typical set up of this apparatus. A hungry rat will be placed into a conditioning chamber (also sometimes called 'a Skinner box', after the psychologist B. F. Skinner who used this sort of apparatus to study learning), and be allowed to investigate the area. During this time the rat will perform many responses, including grooming or standing on its hind legs to investigate the roof of the chamber. However, quite by accident,

6. A laboratory rat pressing a lever in a typical conditioning chamber. In the top right of the picture you can see a tub of food pellets. When the lever is pressed one of these pellets will fall down a tube and be delivered into the food-well to the left of the lever.

it may eventually press the lever. When it does so, the rat will be given a reinforcer, such as a tiny food pellet. At first during the experiment, the psychologist might arrange for every lever press to result in the delivery of food. Eventually, with more training, however, the experimenter may decide that food is delivered only after the rat has made a certain *number* of lever presses, or after a certain period of *time* has passed since the last food pellet was delivered. Introducing these more stringent schedules of reinforcement provides a better model of the sorts of relationships that exist between behaviour and reward in the natural environment, for it is frequently the case that a commodity will only be delivered if a certain amount of effort has been discharged. Rarely is it the case that things work first time, every time.

In some cases, however, the rat may never press the lever by accident—which immediately presents us with a problem, because instrumental conditioning will never take place unless the food reward is delivered. Under these circumstances behaviour must be *shaped*. The trick to this is to reward closer and closer approximations to the desired response. So, for example, the rat might first be rewarded for just being near the lever, and then later for touching the lever or raising its paws above the lever. Eventually these responses will result in the desired behaviour being performed and instrumental conditioning can begin. Shaping is a procedure that people use quite frequently when teaching young children novel actions (such as how to say their first words). It has also been used to help developmentally challenged children to acquire behaviours. For example, Montrose Wolf, Todd Risley, and Hayden Mees used shaping to train an autistic boy to wear his glasses. They did this by rewarding approximations of this behaviour at meal times when the boy was hungry.

Instrumental conditioning can be acquired by pairing what is essentially an arbitrary response—like pressing a lever—with a reward. This observation raises the question of why, then, do people not perform all manner of strange, arbitrary responses more frequently, given that we live in a world full of all manner of potentially rewarding events such as Facebook 'likes', or sugary and fatty fast-food? One answer to this question is to note that, actually, people *do* perform all sorts of arbitrary behaviours. Sportspeople will frequently admit to having a 'lucky' pair of socks (for example) that they will wear quite regularly as a consequence of having once worn them during a victorious sporting event. This sort of behaviour reflects a superstition just as much as not walking under a ladder does—and a similar effect has also been observed in the laboratory pigeon. B. F. Skinner reported an experiment in which pigeons were put into a conditioning chamber where food was delivered to them at regular intervals—irrespective of their behaviour. In six out of eight pigeons Skinner observed idiosyncratic behaviours such as hopping from one leg to the other. Skinner suggested that the way in which these 'superstitious' behaviours were acquired by the pigeons was that they were often first performed, by chance, just prior to the delivery of food, and as a consequence were subsequently repeated through instrumental conditioning.

Although they are relatively common, superstitions do not dominate the behaviour of people in the way that we might expect from an application of the principles of instrumental conditioning to human behaviour. Why might this be so? One reason for why instrumental conditioning might not always be successful is suggested by an experiment conducted by Lynn Hammond. She showed that thirsty rats were far less likely to make a lever press in order to gain access to water if they also received water during the intervals when they were not pressing the lever. Gaining the reward of water, independent of responding to an action, interfered with instrumental conditioning. A similar effect is present in human behaviour: most people are happy to work reasonably hard so long as they receive some reward for doing so (such as a salary). However, if a person's salary were paid irrespective of whether they did any work, there would seem little justification in turning up for work at all.

The spread of learning

Learned behaviour is not only restricted to the stimuli that have been used in training. If a stimulus is conditioned then a response can also be evoked by other similar stimuli, despite the fact that they may never have been explicitly used in producing the trained response. This is called 'stimulus generalization'. Norman Guttman and Harry Kalish provided one of the seminal examples of stimulus generalization. They placed pigeons into a conditioning chamber and rewarded them with grain for pecking at a small key that was illuminated yellow at a wavelength of 580 nanometres (nm). Once the pigeons were pecking frequently at the key light, test trials were introduced. These test trials (which were conducted in the absence of any reward) comprised illuminations of the key at many different wavelengths between and including 520 nm (green) and 640 nm (red). The results indicated that, in addition to pecking very frequently at the wavelength used in training, the pigeons also pecked very frequently at wavelengths of 570 nm and 590 nm—the pigeons generalized their learning from one stimulus to other, rather similar, stimuli. However, this generalization was not indiscriminate—as the wavelength of the test stimuli became more and more different to the wavelength used in training (e.g. at 550 nm and 600 nm), responding became significantly less frequent.

In order to explain how learning may generalize to stimuli that were never used in training, Donald Blough argued that even relatively simple stimuli, like a light, actually contain a collection of features, each of which is conditioned when the light is paired with a reward. Thus, if some of the features of a light that has a wavelength of 580 nm are also present in a light that has a wavelength of 570 nm, then responding will be very frequent to this untrained stimulus—as was observed by Guttman and Kalish. In order to explain why the spread of learning can be restricted—that is to say, why generalization is not indiscriminate—Blough suggested that stimuli that are less similar will share fewer features in common. As a consequence, then, responding to a light that has a wavelength of 550 nm will be relatively weak, as it shares relatively few features with the 580 nm light that was used in training.

This chapter gave a brief, introductory account of learning—in particular in terms of what learning is and how it is studied in animals. Although providing a definition of learning is a thorny affair, quite a lot of what we understand by the term can be understood as 'a relatively permanent change in behaviour as a consequence of experience'. The examples of habituation, and classical and instrumental conditioning all satisfy this definition—albeit in slightly different ways. We will now turn our attention to the question of exactly what is learnt during learning.

Chapter 2
What is learned during learning?

Throughout the study of learning, psychologists have concerned themselves with two key questions: (1) what is actually learned during learning? and (2) what is the nature of the mechanism that allows this learning to take place? This chapter will look at some of the answers that have been provided to the first question whilst the second question will be examined in Chapter 3.

Rather curiously, psychologists have been quite conservative when drawing conclusions about what is learned during their experiments into classical and instrumental conditioning. Recall that Thorndike, who trained cats to escape from an enclosed box by pressing a lever, interpreted what was learned during this training as no more than a connection or an *association* between two events. In Thorndike's case, the association was between a stimulus and a response—a proposal to which we will return in just a moment. Although investigators have suggested that other events may be involved in learning beyond a stimulus and a response, it is still common for them to suggest that the nature of the learning is an association between events. For example, in the case of Pavlov's classical conditioning experiment, psychologists have suggested that an association is formed between the bell (the conditioned stimulus) and food (the unconditioned stimulus) when they are paired.

Why this conservatism? Why just talk of connections between things? Why not, instead, provide a richer explanation for the content of learning? Why not instead suggest that it is a process of reasoning in which the animal deduces that its actions (or, perhaps, some other stimulus in the environment) causes the delivery of, say, food? I think two reasons contribute to this conservatism. First, the organ that is responsible for learning to occur in the first place—the brain—comprises many neurons, each of which connects with many other neurons. For example, the average human brain comprises around eighty billion neurons, each of which may connect to thousands of other neurons. Consequently, the brain is made up of trillions and trillions of connections. 'Connection' is what the brain does really well. Furthermore, the behaviour of these connections can change with experience. In an extremely influential experiment, Timothy Bliss and Terje Lomo stimulated, rather weakly, the neurons in a part of the brain of a rabbit called the hippocampus. They measured activity in neurons downstream of the hippocampus in another region called the dentate gyrus. What they found—not too surprisingly—was that even weak stimulation of the neurons in the hippocampus resulted in a weak response in the downstream dentate gyrus neurons. However, the important discovery made by Bliss and Lomo was that if they electrically stimulated the neurons in the hippocampus more strongly, then later—days later—the connected neurons in the dentate gyrus would respond more strongly even to weak stimulation of the neurons in the hippocampus. In terms of the definition of learning provided in Chapter 1, there has been a relatively long-lasting (i.e. lasting days) change in behaviour (in this case neural behaviour) as a consequence of experience (high intensity electrical stimulation). Thus, conceiving of learning as a change in the connection between things is to propose a mirroring of what happens in the brain—that is to say, the proposal is *biologically plausible*.

The second reason for conservatism when interpreting what is learned during learning is Morgan's canon. Conwy Lloyd Morgan was a British ethologist who studied under 'Darwin's bulldog' Thomas Huxley. Morgan proposed a principle that many comparative psychologists still observe:

> In no case is an animal activity to be interpreted in terms of higher psychological processes if it can be fairly interpreted in terms of processes which stand lower in the scale of psychological evolution and development. (Morgan, *An Introduction to Comparative Psychology*, p. 59)

Morgan's canon tells us that the best explanation for a behaviour is one that refers to the simplest psychological process. It is therefore reminiscent of the more general philosophical and scientific principle, Occam's razor, which today is usually interpreted as 'other things being equal, the simplest explanation is best'. By describing the content of learning as the formation of an association between events—such as an association between a stimulus and a response—psychologists are making very few assumptions about the psychological content of the experience. In fact, the description is going little further than describing the conditions of training themselves. By doing so, theories of learning retain a degree of simplicity about them. This allows other psychologists to conduct experiments to test them relatively easily, and advance our understanding of what is learned during learning.

The items of association

Although psychologists, especially those interested in theories of learning, have been relatively conservative about the process of learning—arguing that it is based upon no more than the acquisition of an association between things—they have been far more liberal with their ideas about what can be associated with what. As we saw in Chapter 1, to which we will return again and again, different types of responses, stimuli, and rewards have been employed as events in classical and instrumental conditioning, and some psychologists have suggested that all of these events are candidates for association with one another. It is therefore instructive to ask whether it is justified to permit this breadth of associative freedom. Why not limit associations, as Thorndike did, to just be between a stimulus and a response? This idea was taken up by a number of North American psychologists such as Clark Hull and Edwin Guthrie in the first half of the 20th century. These so-called *stimulus-response theorists* followed Thorndike's lead, and suggested that both instrumental and classical conditioning can be conceived of as the formation of an association between a stimulus and a response. The role of events like acquisition of food (or water, or sex) is simply to reinforce or strengthen this association. To take the example of the laboratory rat learning to press a lever for food that was described in Chapter 1, the stimulus-response theorists would argue that this behaviour is a consequence of food reinforcing an association between the sight of the lever and the response of pressing it. In the case of Pavlov's experiment, food reinforced the association between the sound of the bell and the response of salivation that was elicited by the food.

There is clearly some virtue in the account of learning that is offered by the stimulus-response theorists. This account allows experience to modify behaviour with a mechanism that has no additional steps between the input of the stimulus and the output of the response; it therefore has merit on the basis of simplicity alone. However, there is also reason to be unsatisfied with the account as well. Is it really the case that motivationally significant events in the world (such as provision of food) serve only to strengthen the association between, for example, the sight of a lever and the response of pressing it? It implies that the animals make no connection between the response and the reinforcer. Thus, after pressing a lever for the umpteenth time, the rat will still be entirely surprised by the presence of food when it is delivered. Could this really be so? The answer, rather confusingly, is sometimes yes—and sometimes no.

Goal-directed actions and habits

According to Thorndike, reward (or 'reinforcement' as it is called in instrumental conditioning) simply acts as a catalyst to drive the association between the stimulus and the response. It follows then that it should not matter if, after instrumental conditioning has taken place, the value of the reward changes—as the reward itself does not form part of the association. To test this prediction, Christopher Adams conducted an experiment in which rats in an experimental group were given the opportunity to make *a hundred* presses of the lever in a Skinner box, each of which resulted in the delivery of a small pellet of sugar as a reward. Once this training was complete, the rats were put through a procedure that was intended to *change the value* of the sugar pellets by establishing a learned flavour aversion to them (see Chapter 1). Thus, in the absence of the levers, the rats ate the sugar pellets, and then sickness was induced (in this case, through an injection of lithium chloride). Once the rats had recovered from the sickness, they were given a final test session in which they could press the lever once more—but in the absence of the delivery of any sugar pellets. The results of this final test session can be seen in the left-hand pair of bars in Figure 7. Relative to rats in a control group (who did not have the sugar pellets paired with sickness), rats in the experimental group made fewer presses of the lever. Changing the value of the reward, then, clearly *did* influence instrumental conditioning.

One interpretation of the results of the experiment by Adams is that learning is goal-directed. That is to say, the animals were making a response in order to gain a valuable outcome—a goal. Humans undoubtedly engage in similar goal-directed learning. For example, upon cooking a meal in the evening, I will make the response of adding salt to the food—not because I reflexively do this when in the presence of certain environmental stimuli, but because I have learned that it results in an improvement in the flavour of the meal. However, it would be a mistake to think that

27

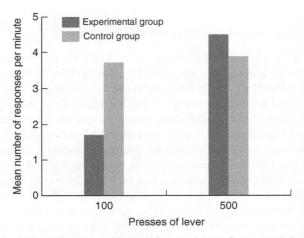

7. **Goal-directed action and habit.** After receiving only some training (100 lever presses), goal devaluation in the experimental group reduced responding relative to a control group who had not had food devalued. After lots of training (500 presses), however, goal devaluation had no effect.

all instrumental behaviour is goal-directed. Adams neatly demonstrated this in rats by including another pair of groups in his experiment who were treated exactly the same as the experimental and control rats described in the earlier paragraph; however, these rats received five times as much instrumental conditioning in the first stage of the experiment as did the other two groups—that is to say, they were given the opportunity to make 500 presses of the lever, each of which resulted in the delivery of a sugar pellet. Rats in the experimental group then had the sugar pellets devalued through flavour aversion training and, finally, both groups were given a test session in which only the lever was presented. The right-hand pair of bars in Figure 7 show the data from this test session. In contrast to the experimental rats who were trained with only 100 lever presses, the experimental group who were trained with 500 lever presses showed no sign of the flavour aversion learning having had an impact on their lever pressing—they continued to press the lever just as frequently as the control group.

In order to explain the results of his experiment, Adams suggested that instrumental conditioning first results in learning that is goal-directed—actions are performed in order to obtain a reinforcer, a response–reinforcer association. Consequently, should the value of the reinforcer change (in this case by being paired with sickness) then it will be reflected in the frequency with which the action is performed. However, with extended training, there is a transition from goal-directed behaviour to a more habit based stimulus-response behaviour that is under the control of environmental stimuli and less to do with the current value of the reinforcer. You might have experienced a similar transition from goal-directed behaviour to habit yourself. Consider the circumstances under which you wipe your feet when you enter your home. In all likelihood, you learned to do this as a child when you were told off by one or both of your parents for not wiping your feet when you had mud on your shoes—or perhaps because you were praised for remembering to do it when walking through the door. Many years later, you will no doubt still find yourself wiping your feet upon entering your home. Do you do this in order to avoid a telling-off, or in order to seek praise? I certainly don't—mainly because I no longer live with the person who taught me to wipe my feet in the first place, in fact I sometimes wipe my feet even when my shoes are not dirty, or if I am going to immediately take off my shoes upon entering the house. The simple act of wiping your feet upon entering your house—many times—seems to have eventually resulted in the stimuli around the door (such as the sight of the door mat or coats hanging near the front door) becoming associated with the response of feet wiping: a habit has been acquired. As we shall see later in this book, the notion of habitual responding has been particularly important to our understanding of other behaviour, such as drug use. But as far as the current discussion is concerned, the results of Christopher Adams's studies show that Thorndike's proposal may not be entirely without merit.

The content of classical conditioning

Although Thorndike's suggestion about the content of learning during instrumental conditioning is, in some ways, counterintuitive, the same might not be said for classical conditioning. When people talk about their behaviour being 'Pavlovian' or 'conditioned', they often mean that they are behaving in a rather reflexive manner—as if the conditioned response is a mindless automatic consequence of the presentation of the conditioned stimulus. However, this belief is not strictly correct, as the unconditioned stimulus in classical conditioning can form part of the learning experience in the same manner as the reinforcer does in instrumental conditioning. This was demonstrated in an experiment by Ruth Colwill and Daphne Motzkin. They used a procedure that was similar to that used by Adams, but instead of investigating what was learned during instrumental conditioning they investigated the content of classical conditioning. Rats first received trials in which a tone was presented for ten seconds. Following each tone, food was delivered to the rat. After a number of days of this training, the rats had the food devalued in the same manner as in the experiments by Adams—the consumption of the food pellets was paired with drug-induced sickness. In a final test with the tone (in the absence of food), Colwill and Motzkin showed that conditioned responding was significantly reduced. This result implies that the effect of the unconditioned stimulus (in this case the provision of food) in classical conditioning is not just as a catalyst to forging an association between the tone and a response (such as retrieving the food)—a view which would be in keeping with the view that classical conditioning is only automatic and reflexive. Instead the association is between the stimuli—the conditioned and unconditioned stimuli. This implies that conditioned responding reflects, at least in part, the animal's knowledge about what stimuli are likely to occur in the near future.

Imaginary conditioning

The results of the experiment by Colwill and Motzkin could also be taken to suggest that conditioning results in one stimulus calling to mind some form of representation, or image, of its associate. This follows because, during the test, the conditioned response to the tone was a consequence of its association with food—and yet food was not present at the test, and so some form of image of the food was presumably influencing behaviour. This analysis has some appeal: as a young PhD student I can recall being in the laboratory, struggling to load up an archaic printer with paper, whilst at the same time thinking hard about the complexities of a theory of learning proposed by a psychologist called Nick Mackintosh. I must have been doing both of these tasks for some time, or with a degree of concentration, as they left an impression upon me; for subsequently, whenever I had to load the printer with paper I ended up thinking about Mackintosh—even though, on these later occasions, I hadn't actually planned to think about his theory before approaching the printer. I am sure that you will have also encountered times when something has been called to mind simply by being in a situation or context where that something has previously happened—people's ability to better recall previously learned items when placed back into the earlier context of learning is a classic example of such an effect. However, the example I have just described—of the printer and the psychologist—is a little more unusual, for it describes a situation in which I made an association between something that was physically present (a printer) and something that was present only in my mind. Nick Mackintosh wasn't actually in the lab with me whilst I was grappling with the printer but, nonetheless, I still developed an association between him and the printer.

It turns out that associating imaginary items with those physically present is not only limited to humans, animals also show evidence of doing the same thing. Peter Holland and his colleagues conducted a series of experiments showing that laboratory rats show a similar effect—that an association can be acquired between a stimulus that is actually present and one that is not. In the first stage of one of Holland's experiments a tone (the conditioned stimulus) was paired with food (the unconditioned stimulus). Once this training was complete, the rats received trials in which the tone was paired with a different, aversive, unconditioned stimulus—lithium chloride—which you will recall has the effect of making the rat feel sick. Interestingly, Holland observed that this training resulted in the rats subsequently avoiding the food that the tone had been paired with in the first stage of the experiment—even though the food itself had never been paired with any unpleasant event. The way that Peter Holland interpreted this (and many other accompanying studies) is that in the first stage of the experiment, pairing the tone with food meant that the rat came to imagine food when that tone was played—just as I would imagine Nick Mackintosh when I was in front of that printer. Consequently, in the second stage of the experiment, when the tone (previously associated with food) was paired with nausea, the rat would be imagining the food at the time at which it was made ill. This would allow the two things, which otherwise had never been presented together (food and nausea), to be associated, and this consequently led to the formation of an aversion to the food.

Imaginary conditioning is not limited only to circumstances in which the associate of the imagined event is physically present at that time. Studies have now shown that animals can associate two events even when they are *both* imagined. Dominic Dwyer, Nick Mackintosh, and Bob Boakes reported a series of experiments that demonstrated just this. The details of these experiments are rather complex, but the general idea is quite simple (see Figure 8). First, Dwyer and colleagues paired two events with one another: (a) a distinctive context and (b) the odour of peppermint; and then on a separate occasion another two events with each other: (c) the odour of almond with (d) a drink of sucrose. In the second stage of the experiment the rats were placed into (a) the distinctive

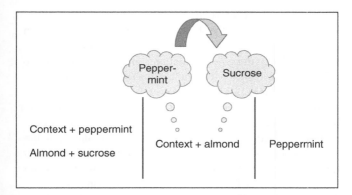

8. Associations between imaginary events. The three stages of the experiment reported by Dwyer, Mackintosh, and Boakes. Rats imagined peppermint and sucrose (shown in thought bubbles) at about the same time, resulting in their association.

context and exposed to (c) the odour of almond. The idea behind their experiment was that in the second stage, exposure to the distinctive context and the odour of almond at the same time would result in the rat also imagining the associates of these two events—peppermint and sucrose—at the same time. As rats really quite like sucrose, this should mean that they should learn to acquire a preference for peppermint, as it is paired with this tasty treat—albeit, all in the mind. This is precisely the result that Dwyer and his colleagues observed: at the end of the experiment, when peppermint was given to the rats on its own, they showed an increased preference for it.

Learning about things that don't matter

The experiments described so far have provided some quite impressive feats of cognition from a creature as humble as the laboratory rat. We have certainly come quite a long way from Thorndike's idea that all that is learned during learning is a rather mindless connection between no more than the input and the output of the learning system—a stimulus and a response. Instead, we now have evidence that this animal seems to be in possession of something akin to private, internal, representations, and furthermore that these representations can be linked together to influence behaviour. What is notable in all of the experiments described so far in this book, however, is that an event of motivational significance is used as one of the items of association: be it something pleasant like food, water, or sex; or something aversive like pain or sickness. At one level, it is understandable why these sorts of events have to be used in studies of animal learning—the definition of learning provided in Chapter 1 requires us to measure some form of behaviour, and behaviour is more likely to be observable when an event of motivational significance is involved.

However, does learning require the use of motivationally significant events? Probably not. We already saw in the experiment by Dwyer, Mackintosh, and Boakes that rats could associate a distinctive context (which in this experiment was a bucket) with an odour. If they had not associated these two, relatively unimportant stimuli there would be no basis upon which the whole experiment would work—it is a necessary link in the chain of logic that holds the whole interpretation of the experiment together. It is important, though, to demonstrate the generality of this effect—to show that organisms *are* capable of forming associations between motivationally neutral events. This is because theories of associative learning have been used by some as a more general-purpose tool for explaining how organisms learn about the structure of their environment—irrespective of whether events of importance are involved. If associative learning is only limited to circumstances when motivationally significant events are involved, this approach would be undermined. Thankfully, it is clear that animals are pretty good at associating events, even when the events are relatively boring. Robert Rescorla and Paula Durlach conducted an elegant experiment with pigeons to demonstrate that this was so. In the first stage of this experiment, the pigeons were exposed to trials in which an image of a blue semi-circle (shown here in black) was presented next to an image of some lines angled at 45 degrees from vertical. On other trials the pigeons saw an image of a yellow semi-circle (shown here in grey) next to an image of lines that were this time angled at –45 degrees from vertical. The purpose of this training was to provide the pigeons with an opportunity to associate the different colours with the different orientations of the lines. The problem, however, is how to reveal this learning. The way in which Rescorla and Durlach solved this problem was to subsequently give the pigeons a second stage of training in which they established a conditioned response to the 45-degree lines by pairing them with access to food, but not with the lines that were –45 degrees. If the pigeons had associated blue (here, black) with 45-degree lines and yellow (here, grey) with –45-degree lines in the first stage of the experiment then changing the value of the line orientations should also affect responding to the colours that they were associated with. This was exactly what Rescorla and Durlach observed: they gave the birds a final test with the blue and yellow colours alone (see Figure 9) and found that the pigeons' rate of responding (pecking at the images) was more rapid to blue than to yellow.

Rescorla and Durlach's experiment is a master class in experimental design, as each stimulus, either the crucial test stimuli or their associates, is presented equally frequently, and

9. **Sensory pre-conditioning. The three stages of the experiment reported by Rescorla and Durlach. Pigeons associated different shades with different line orientations before one of the orientations was conditioned in stage two. This resulted in a greater mean number of responses per minute (RPM) to black than to grey.**

furthermore the pigeon serves as its own control (there is no need for a control group in this experiment). However, beyond such admiration, this experiment demonstrates that pigeons, just by being exposed to two relatively neutral images at the same time (a colour and a set of lines), can associate these images without any motivation for doing so, supporting the idea that an association can be acquired between events that contain little (or no) intrinsic motivational significance for the organism.

The experiments in this chapter have introduced the content of learning during instrumental and classical conditioning. It is, of course, impossible in a book of this size to describe the many other intriguing things that we now know that animals can learn about. We have seen that actions performed during instrumental conditioning can be either goal-directed or habitual—depending on how well advanced learning is. We have also noted that classical conditioning is not as reflexive as one might first imagine from its description in popular culture. Classical conditioning, through the process of associative learning, serves to provide the organism with a way of predicting its future—both when the future is important and when it is not. We have therefore provided some answers to the first of the two key questions that can be asked about learning: what is actually learned during learning? The purpose of Chapter 3 is to answer the second question: what is the nature of the mechanism that allows this learning to take place?

Chapter 3
The surprising thing about learning

Let us start with a thought experiment. Imagine you have just graduated and are starting your first job. You form a partnership with a businessman (let's call him William) who, year in, year out, for the past ten years, has made a steady profit of £500,000. After working with William for two years you get a chance to look at the accounts of the company. What you discover is that, like the past ten years, the company has continued to make a steady profit of £500,000. Who do you think is responsible for the success of the company over the last two years? Is it you or William?

A moment's reflection might lead to the conclusion that, unfortunately, you've had rather little impact upon the profits of the company. Although the company has made a healthy, steady profit over the two years that you have worked there, the company did exactly the same every year for the ten years that William was working on his own. So, despite your best efforts, it seems as if you made rather little impact upon the success of the company. Now, the outcome of this thought experiment might seem, perhaps, a little unfair to you. After all, in this imaginary scenario you did actually work for two years in a company that made a profit. Fairness aside, however, this result—*not attributing company profit to you because of William's past success*—is an example of one of the most influential phenomena to have been discovered about learning—a phenomenon that is present across the animal kingdom, from honeybees, goldfish, and rats to monkeys and even rather more intelligent creatures such as university undergraduates. This phenomenon is called 'blocking'.

Blocking

Leon Kamin provided the classic demonstration of blocking in 1968. He conducted a fear conditioning experiment, the details of which are very dissimilar to the thought experiment that we have just conducted. The structures of the thought experiment and Kamin's experiment are, however, comparable, as we shall see. In the first stage of Kamin's experiment a group of rats was given sixteen trials in which a burst of white noise was played for three minutes. At the end of each trial a mild electric shock was applied to the feet of the rats. After this stage of the experiment the rats received a further eight trials in which the burst of white noise was again followed by the mild electric shock. However, in this stage of the experiment a light was turned on at the same time as the noise. At the end of the experiment Kamin turned on just the light—and, in the absence of the noise (and any shocks), he measured how much fear the rats displayed (see Figure 10). The results were rather striking. Despite the fact that the light had been paired with an electric shock eight times, the rats showed no more fear when the light was turned on than when it was turned off. It appeared to Kamin as if the rats hadn't learned anything about the light at all.

Now, you might argue perhaps that the rats just weren't very good at associating the light with the shock or that, for some peculiar innate reason, the rats quite liked the sight of the light, and this prevented them from displaying any fear to it. Fortunately, Kamin was able to rule out both of these possibilities, as he also included in his experiment a control group of rats. The control group was treated in an identical fashion to the first group, with one important exception: they didn't receive the first stage of training in which the noise was paired with the shock. They only received

	Stage 1	Stage 2	Test
Blocking group	Noise → Shock	Noise + Light → Shock	Light (no fear)
Control group		Noise + Light → Shock	Light (lots of fear!)

10. **An overview of the design of Leon Kamin's blocking experiment.**

the second stage of training in which the light *and* the noise, presented together, were paired with the shock. When Kamin presented the light to these rats at the end of the experiment the rats demonstrated substantial fear.

What are we to make of Kamin's experiment? I would like to emphasize three points. First, Kamin's blocking experiment has been successfully replicated, time and time again, in many laboratories across the world, using a variety of different methods of studying learning, in all sorts of different species—blocking, thus, seems to be a general feature of learning that pervades the animal kingdom. Second, there is a similarity between Kamin's experiment and our earlier thought experiment: Kamin showed that his rats did not associate the *light* with the shock when it was accompanied by a *noise* that had, in the past, been associated, alone, with shock. In the thought experiment, *you* were not associated with company profits because you were accompanied by *William* who had, in the past, been associated with company profit. Third, psychologists have suggested that the reason why the association between the light and the shock (or between you and the profit) was not formed was because at the time at which these events co-occurred *there was no surprise*. So let us now consider the factor of surprise.

What is surprise?

Two things probably come to mind when you are asked what surprise is, one will probably be the expression of astonishment and the other the experience of some kind of a sensation—a jolt, perhaps. Although these two things often accompany surprise, they are perhaps better described as responses to surprise rather than surprise itself. Consequently, psychologists have attempted to provide a definition of what surprise is. Inspired by Kamin's blocking experiment, Robert Rescorla and Allan Wagner suggested in 1972 that we can think of surprise as a violation of an expectation. For example, you might expect your favourite football team currently languishing at the bottom of the division to do badly in their next match, perhaps because they have lost every game that they have played for the past six weeks. If your team should however go on to win their next match, you will be (pleasantly) surprised. Why? Because your expectations have been violated: you expected a loss, but what actually happened was that your team won. In other words, *there was a difference between what you expected to happen and what actually happened*. This statement, in italics, summarizes, perhaps, one of the most important concepts in learning. It is a definition of surprise that, as we shall see, is thought to be at the heart of learning. Another advantage of this statement is that we can express it in the form of a simple mathematical equation:

Surprise = What happened – What you expected to happen.

From this equation, it should be clear that the greater the difference between *what happened* and *what you expected to happen*, the greater the surprise. And, conversely, if there is no difference between *what happened* and *what you expected to happen*, there will be no surprise. If we now go one step further, as Rescorla and Wagner did, and propose that the amount of learning that takes place is related to how surprised the animal is

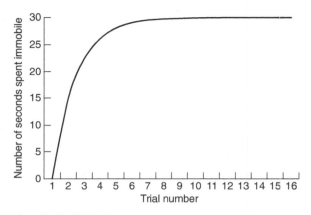

11. A hypothetical learning curve.

(that is to say, Learning ≈ Surprise), two things will follow. First, we have a way of understanding why learning slows down and eventually stops; and, second, we can start to make sense of Kamin's experiment. I shall deal with these two issues in turn.

We can use the concept of surprise to determine how learning might proceed across trials. Consider the graph in Figure 11, which portrays the results of an imaginary experiment in which, on each of sixteen conditioning trials, a rat is played a noise that is followed soon after by the delivery of a mild electric shock. The vertical axis on this graph is a measure of how much the rat has learned, as measured by how much time it spends remaining immobile (in fear) during the noise. We can see that, early on during this imaginary experiment, there is a rapid increase in the amount of time that the rat is immobile—the learning curve is steep. This is because there is a large difference between what the rat expects to happen during the noise (i.e. not much really) and what actually does happen (an electric shock). The rat is therefore very surprised and learns quickly. As this learning starts to accumulate, however, the difference between what the rat expects to happen (something painful) and what actually does happen (an electric shock) gets smaller and smaller. This reduction in surprise slows down the speed of learning until, ultimately, it reaches a point at which the rat is fully expecting the shock to occur after the noise, and learning no longer takes place—and the learning curve flattens off.

This hypothetical learning curve matches the behaviour of animals too: if Figure 4(a) is consulted once again, we can see that most of the learning took place during the first five days of training; after this, there was relatively little change in behaviour.

Making sense of Kamin's experiment

So, imagine now that after giving the sixteen conditioning trials with the noise and the shock we do what Leon Kamin did in 1968 and turn on a light at the same time as the noise. If we give the rat a shock after the noise and the light, should it learn? On the basis of what we have just considered the answer should be 'no'. The difference between what the rat *expects to happen* and *what does happen* is about zero; so it won't be surprised—and according to Rescorla and Wagner no learning will take place. This means that the rat will not learn anything more about the noise, but more importantly it won't learn anything about the light. So if you were to present the light to the rat at the end of the experiment, without the noise or the shock, as Kamin did, the rat would not fear it. We can say that the noise blocked learning about the light. To return to our thought experiment from earlier, it should also be apparent why the profits of William's company were not attributed to you. As a consequence of the ten years of success and profit with William, it is to be expected that the company will make a profit in the future. As a consequence, it is not at all surprising when the company continues to make a profit when you join it. Therefore your contribution is not learned about. You have been blocked!

How to stop blocking

So what can be done to combat this state of affairs? What should be done, in our thought experiment, to demonstrate that you are having an effect on the company profits—how can you stop blocking from taking place? A moment's reflection on the proposal put forward by Rescorla and Wagner provides one possible answer: you need to make a difference. The reason why company profits were not attributed to you—despite the fact that you were employed during a time of profit—was because what actually happened (the company made a profit) matched expectations (the company will make a profit). Only when an expectation is violated will learning take place. The trick, then, is to make sure expectations are exceeded—and one way of doing this would be to ensure company profits go up when you are employed. For example, if when you join William's company profits go up by £100,000, then there will be a difference of £100,000 between what is expected (a profit of £500,000) and what is obtained (£600,000), and learning should be able to once again take place. You will be associated with company profit (and hopefully suitably rewarded). And you will have been unblocked.

It turns out that this 'unblocking' effect is also evident in non-human animals. Recall that in Kamin's original blocking experiment he gave rats pairings of a noise with a mild electric footshock, before giving them presentations of the noise at the same time as a light with the same mild electric footshock. His results revealed that the rats learned very little about the light: when it was presented alone at the end of the experiment they expressed little fear in response to it. However, in a second experiment, Kamin increased the strength of the shock when the light was added to the noise in the second stage. Under these circumstances the rats showed significant learning about the association between the light and the shock. They too experienced an unblocking effect. The explanation for Kamin's second result is similar to that for our analysis of William's company. The rat was expecting one thing on the basis of the noise (a mild shock) but then received something rather different when the light was added to the noise (a stronger shock). This difference between expectation and reality was a surprise, which permitted learning to take place—with fear being acquired in response to the light.

Surprise by omission

So far in this chapter we have limited our consideration of the nature of surprise to when it is (a) positive—where what happens *is greater than* expected; or (b) zero—where what happens *matches* expectation. However, a third alternative is clearly also possible: (c) negative—where what happens *is less than* expected. We have already encountered this third possibility in Chapter 1, when we discussed extinction. Recall that extinction refers to a situation when a learned behaviour, first established by pairing a conditioned stimulus with an unconditioned stimulus, is weakened by repeatedly presenting the conditioned stimulus alone. Extinction constitutes exactly the kind of circumstance in which what happens is less than our expectation: an unconditioned stimulus is expected, but then not delivered. According to Rescorla and Wagner's explanation of learning, there is a violation of expectation here, so learning should take place. But, rather than the learning being positive, it is negative—so rather than the connection between the conditioned and unconditioned stimuli being strengthened, it is therefore *weakened*, and, as a consequence, conditioned responding will get correspondingly weaker.

Interestingly, and perhaps counterintuitively, the connection between the conditioned and unconditioned stimuli can also be weakened if the unconditioned stimulus is presented during learning—this is an effect called overexpectation. Imagine the situation in which two new employees, Alice and Barry, join William's company, and they each have a history of being responsible for making other companies increase their profit by around £50,000. After being with the company for a year it is discovered that company profits did indeed increase, but only by £50,000. In this situation there is an expectation of profits increasing by £100,000—as Alice and Barry should *each* generate a £50,000 increase in company profits—rather than the £50,000 actually generated overall. There is therefore a case of overexpectation here—much is expected, but little is achieved, which, as in our previous example of extinction, results in surprise by omission. According to the theory proposed by Rescorla and Wagner, this sort of situation should result in Alice and Barry being associated with a reduction in company profit. This is counterintuitive because Alice and Barry were both paired with increased company profits. Sadly for them, they were not paired with enough increased profit.

This sort of overexpectation effect is also observed in studies of animal learning. In an experiment reported by Matthew Lattal and Sadahiko Nakajima, rats were given trials in which a light, a noise, and a tone were all individually paired with a single pellet of food. This training resulted in the successful acquisition of conditioned responding because, by the end of this stage of the experiment, the rats explored the food-well during the three stimuli more than they did at the start of the stage. Once this training was complete, the light and the noise were presented to the rat at the same time, and were once again paired with a single food pellet. Perversely, according to Rescorla and Wagner's theory, this sort of training should result in the association between the noise and food *weakening*. This follows because, on the basis of the training in the first stage of the experiment, there is an expectation of two food pellets when the light and the noise come on together, but when only one food pellet is received there is surprise by omission, which means that conditioned responding should reduce. This is precisely what Lattal and Nakajima observed. They gave the rats test trials at the end of the experiment in which the noise and the tone were each played on their own. They found that responding was stronger to the tone than to the noise—despite the fact that previously the noise had been paired with food more frequently than had the tone.

So far within this section of the chapter we have considered situations in which a surprise by omission occurs at a time when a conditioned stimulus is presented having already been established as a predictor of an unconditioned stimulus—for example, during extinction, when the unconditioned stimulus is surprisingly omitted after many trials in which it, and the conditioned stimulus, have been paired. However, what would happen if surprise by omission occurs at a time when a conditioned stimulus is presented that has never been paired with an event like an unconditioned stimulus—and which is, in terms of conditioning, neutral? Let us return to our example of you joining William's company. Imagine, at the end of your first year with the company (which, you will recall, usually makes a profit of £500,000 a year) it only breaks even, failing to make a profit for the first time in its history. From past experience a healthy profit should have been made, but when you join, profits are absent. It appears that you are, in some way, preventing profitability. A stimulus that prevents an event, such as an unconditioned stimulus, from happening is known as a 'conditioned inhibitor', and was first discovered by Pavlov himself.

Many studies have now confirmed Pavlov's discovery of conditioned inhibition. Charles Zimmer-Hart and Robert Rescorla, for example, reported an experiment in which rats first received trials where the sound of a clicker was paired with a mild footshock before later intermixing these trials with other trials in which a compound of the clicker and a light was followed by no shock. In this experiment, then, the light serves to inform the rat of the absence of the shock, which, as in the example in the previous paragraph, should establish this stimulus as a conditioned inhibitor. However, this immediately presents a problem: how would one detect a conditioned inhibitor? It is relatively straightforward to detect learning about a stimulus that signals the presence of an event, such as a shock, as the stimulus elicits a conditioned response. But detecting that a stimulus signals the absence of an event is rather more difficult as there will be no response to detect. To overcome this problem, Zimmer-Hart and Rescorla presented the light in compound with another stimulus, a tone, that had been separately paired with shock. The logic here is that if the light has the properties of a conditioned inhibitor—that is to say it signals the absence of shock—then it should interfere with conditioned responding to the tone. This was precisely the result observed by Zimmer-Hart and Rescorla: conditioned responding to the tone was weaker when it was played along with the light.

Problems, problems

The theory of learning proposed by Rescorla and Wagner has been extremely successful. Their explanation for how organisms learn has been applied to a diverse range of situations (as we shall see in Chapters 4 and 5) and has even been applied to our understanding of how dopamine neurons in the midbrain respond to stimuli that are paired with rewards. However, their explanation for learning has a number of known problems.

The first problem for the theory is that surprise by omission does not only weaken associations. If it did, then the only way of re-establishing conditioned responding after, for example, extinction training would be to pair the stimulus with the unconditioned stimulus once again. However, this is not the case: extinction seems to be a more fragile effect than does acquisition. For example, as we saw with spontaneous recovery in Figure 4(b), simply allowing time to pass after extinction is enough for responding to recover.

The second problem with Rescorla and Wagner's theory is that it doesn't explain behaviour that people have little trouble acquiring—so called *configural learning*. For example, I have learned that orange juice, when drunk alone, has a nice flavour, and also that toothpaste, when brushing my teeth, tastes quite nice. However, the combination of the two is unpleasant (try drinking orange juice just after brushing your teeth if you are unconvinced). Rescorla and Wagner's theory doesn't explain why, in this case, I avoid drinking orange juice after brushing my teeth, because the theory treats environmental events in an elemental fashion. That is to say, the theory assumes that the response to a compound of two stimuli will be determined by the sum of what has been learned about each of them. According to the theory, if I learn that toothpaste and orange juice each taste good, then I will regard the two of them together as being very tasty. The theory does not provide the capacity for us to learn that, for example, orange juice is nice when it is consumed alone, but unpleasant when consumed with (or shortly after) the use of toothpaste.

In order to account for how it is possible to learn that the same stimulus can signal different outcomes in different circumstances, our theories of learning have had to be modified. One proposal to this end came from John Pearce. He suggested that organisms associate the entire *combination* of various stimuli with a particular outcome during learning. Thus, rather than treating orange juice and toothpaste as separate elements, Pearce argues that the conjunction of orange juice and toothpaste would be, effectively, perceived as being a single new entity, qualitatively different to the simple sum of orange juice and toothpaste. By treating stimuli, and their combinations, in this manner, Pearce's theory of learning has enjoyed considerable success, explaining how animals solve complex problems that pose a serious challenge to Rescorla and Wagner's theory.

How does learning affect stimulus representations?

The final issue that I would like us to consider in this chapter is how learning might change the way in which events in the world are represented. If, for example, food is delivered to a rat every time that a light is turned on, will the representations of the light and the food change as a consequence of this training? Does learning alter the way in which we perceive the world? The answer to this question is 'yes'. It is now well established that the representations of stimuli change as a consequence of predicting other stimuli, or themselves being predicted. Gregory Kimble and John Ost investigated how learning influenced people's responses to a puff of air directed at the eye. In this experiment, as in the experiment by Martin and Levey that was described in Chapter 1, people received trials in which the illumination of a light was paired, fifty times, with a brief puff of air to the participant's eye, which caused them to blink. Interestingly, as this training proceeded, the eye-blink response to the puff of air diminished. As the association between the light and the air puff grew, the effectiveness of the air puff, it seems, diminished. To confirm this conclusion, Kimble and Ost conducted a test trial at the end of their experiment in which the puff of air was presented in the absence of the light. This test revealed that the eye-blink response to the puff of air was restored. It seems that *an expected event is in some way less effective—or perhaps less salient—than an unexpected event*. This statement appears to have some generality: a similar loss of effectiveness of an expected event has also been observed in eye-blink conditioning in rabbits. Furthermore, it also appears to hold true for relatively innocuous stimuli. In an experiment conducted by Rob Honey and his colleagues, rats were given trials in which the sounding of a tone signalled that a light on the left-hand side of a training chamber would be illuminated, and the sounding of a click signalled that a light on the right-hand side of the chamber would be illuminated. As these trials progressed, the rats spent progressively less time looking at the lights—the effectiveness of each light diminished as training progressed. In an elegant final test, Honey and colleagues switched the auditory and visual stimuli, so that the tone preceded the right light and the click preceded the left light. Under these circumstances, the rats reverted to looking at the lights—when

The surprising thing about learning

49

the light was unexpected it was found to be more effective at attracting the rats' attention.

Attention to a stimulus can also be modified as a consequence of it being established as a predictor of a subsequent stimulus. A simple demonstration of this was reported by Peter Holland in 1977, who, like Honey and colleagues, showed that repeatedly turning on a light resulted in a decline in the extent to which the rats looked at it. However, if the light was subsequently established as a predictor of food, then the extent to which the rats looked at the light increased. Establishing a cue as a predictor of some other event, it seems, can have the result of increasing its capacity to capture attention. In fact it is now well recognized across a variety of species—including humans—that *attention can be captured by stimuli which have been already established as good predictors of a subsequent event*. Experiments conducted by Mike Le Pelley and his colleagues have shown this finding particularly clearly. They presented human participants with pairs of nonsense words on the screen of a computer (for example, 'Jominoid and Dusapplity') and participants had to predict whether these word pairs would be followed by either the sound of a squeak or a 'boing'. There were many pairs of words in the experiment, but importantly only half of them were actually predictive of which kind of noise would be played, the remainder were irrelevant to the solution of the task. Using an eye-tracking procedure, Le Pelley and his colleagues discovered that participants spent more time looking at the predictive words than the irrelevant words. Rather like Peter Holland's rats, then, Mike Le Pelley and colleagues' humans had their attention captured by a cue that was established as a good predictor of a subsequent event, even a rather boring one.

On the one hand, the results just described make a lot of sense. Learning has evolved in such a manner as to help us filter out irrelevant information, and to permit us to focus instead on stimuli that are relevant to a particular task—whether that be finding food for a hungry rat or solving a computer based problem in humans. On the other hand, however, a number of authors have suggested that this kind of relationship between learning and attention is counterintuitive. It makes little sense to devote attention to a stimulus once you have established it as a predictor of an outcome—by this time, the task is solved. Instead, it would make more sense to devote attention to stimuli whose predictive significance is uncertain, in an attempt to ease the acquisition of information. That is to say, it makes more sense *to pay attention to stimuli that are associated with surprise.* Rather confusingly, there is good evidence to support this statement too. Experiments conducted with both rats and humans have shown that both will spend more time looking at a stimulus that is only occasionally followed by an event (thus sustaining an element of surprise) than a stimulus that is consistently followed by an event (which ultimately has become entirely predictable).

The contrasting results of the effects of making a stimulus either a good predictor of a subsequent event or an uncertain predictor of an event has led some psychologists to conclude that we need theories of learning that affect attention in different ways. Only future research will be able to tell whether this conclusion is valid or not. For the time being, however, what seems more certain is that learning does influence stimulus representations, and that it may do this by changing their capacity to attract attention.

The role of surprise has been particularly influential in the study of learning. As we have seen, this simple principle allows learning to be selective—organisms will not blindly associate any events that happen to co-occur, as is so often assumed of classical conditioning. Instead animals restrict their learning to circumstances in which there is a difference between what is expected and what is actually obtained. The way this appears to work, at a psychological level, is by changing the effectiveness (or salience) of stimuli (however one might wish to represent them). Thus an event loses its effectiveness to attract attention as it becomes well predicted by a preceding stimulus. At the same time stimuli come to vary in the extent to which they can attract attention—seemingly to two ends: (1) to assist in choosing relevant information and to tune out irrelevance; and (2) to highlight uncertainty in the world. How these different observations interact with one another is a question that a number of psychologists are currently trying to work out.

Chapter 4
Learning about space and time

Chapters 1 to 3 have described what learning is, what some of its contents are, and the mechanisms that permit it to happen. We have seen how a relatively simple concept—association—provides a good explanation for a number of features of learning across a variety of species and learning procedures. In the case of classical conditioning, it permits organisms to anticipate the world which they inhabit—to respond appropriately and, in the case of instrumental conditioning, to take control of their environment for material gains. However, without two key pieces of information this learning is, largely, useless. Consider the case of a detective who has associated the clues left behind by a jewel thief during his heists. From this, the detective may be able to predict that another heist will happen. These clues will not help our detective, however, unless she knows *where* the jewel thief is going to strike and *when*. The same logic applies to instances of learning in the animal kingdom. Take classical conditioning, for example. It is of little use to a rat if some aspect of the environment provides a signal for food if it does not also tell the rat *where* the food is going to be and *when*, so that the rat can be in the right place at the right time in order to exploit this learning.

In this chapter, I will describe how organisms learn about when and where things are going to happen. As will be seen, both of these aspects of learning have alternative explanations. Some psychologists have tried to explain space and time in terms of associations. However, other psychologists have attempted to explain timing and spatial learning by employing psychological versions of the things that allow us to time things and navigate in the real world—clocks and maps.

Time

Timing has been a key part of the study of learning for many years—in the investigation of both classical and instrumental conditioning. If an unconditioned stimulus (e.g. a shock) *immediately* follows a conditioned stimulus (e.g. a light) then conditioned responding will be acquired more successfully than if a delay of several seconds is placed between these two stimuli. Similarly, if a reinforcer (e.g. food) immediately follows a response (e.g. a lever press) then instrumental conditioning will be much more successfully acquired than if a few seconds is placed between these two events.

One way in which timing can be studied in animals is by using a technique called the *peak procedure*. Here animals are rewarded after making a response during a stimulus after a set period of time has elapsed. Once responding has been established, test trials (so-called 'peak' trials) are given in which the stimulus is presented for a protracted period of time (often far longer than during training), but no reward is delivered, so that responding can be observed without contamination from any responses that may be produced by the delivery of food itself (e.g. sitting back and eating it). William Roberts reported a peak procedure experiment in which a light was turned on and a small lever inserted into a conditioning chamber. The rats could press the lever at any time during the light, but only after twenty seconds would it result in the delivery of food. Occasionally the rats were given peak test trials in which the stimulus was presented for at

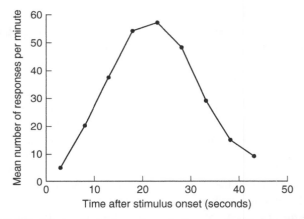

12. **Interval timing. Animals can learn to time events quite accurately. Here rats were trained that food could be earned 20 seconds after a light was illuminated if they made a response at this time.**

least forty seconds and the rats' responses were without effect. The results of Roberts's peak trials are shown in Figure 12. We can see that as the duration of the stimulus increased, the rate of responding increased, peaking at around twenty seconds before then reducing again.

The peak procedure has now been used in many experiments by many different researchers, using a variety of different species, stimuli, and durations, and it provides good evidence that animals are capable of producing an estimate of when events occur (e.g. the opportunity to access food). How might animals perform this feat of timing?

A clock theory of timing

A number of psychologists, such as John Gibbon and Russell Church, have proposed that timing is achieved with a system that, in part, resembles the workings of a clock (plus some memory and a decision making process). Figure 13 shows how the system could work. A *pacemaker* provides an input to the system (you can think of it as the pendulum on an old grandfather clock). This provides a constant stream of pulses. If a stimulus is presented to the animal (e.g. a light is turned on) a switch is opened that allows these pulses to pass into a *counter*, which stores the accumulating pulses, and which you can think of as the current short-term memory representation of the elapsed interval that the animal is trying to time. Should a second signal be presented to the animal (e.g. the light being turned off and food being delivered) then the switch is closed and the number of pulses in the counter is passed into long-term memory as the *rewarded duration*. Should the light be presented to the animal again then pulses will once again feed into the counter, but now the value stored in the counter (short-term memory) is compared to the stored duration in long-term memory. When these two durations are sufficiently similar, the system produces *a decision* about whether to respond.

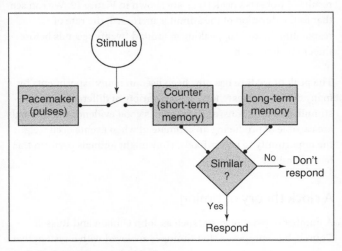

13. **A clock-based system of timing.**

The clock based system of timing has a degree of intuitive appeal. It also provides a straightforward explanation for the results of the peak timing experiment conducted by Roberts that was described earlier. At the onset of the light the pulses from the pacemaker will feed into short-term memory and be counted. When lever pressing by the rat is rewarded with food and the light is turned off, the duration in short-term memory will be sent to long-term memory. On peak trials the rat will continually be making a decision about whether the current duration in short-term memory is similar to that stored in long-term memory. If it is, then the rat will respond. If it is much shorter, or much longer, then the rat will not—which is precisely what was observed by Roberts.

One problem that clocks have is that their accuracy depends upon the pacemaker being constant. If the pendulum of a grandfather clock is shortened, for example, then the clock will start to run too fast. A similar effect has also been observed in animal timing. If rats are given methamphetamine (or 'speed') after having learned a timing task without the drug, then they will behave as if the timed intervals are longer than they really are—a behaviour that is consistent with the idea that the rate of pulses emanating from the pacemaker has speeded up.

One system of timing or two?

The clock based system has been very influential in the study of interval timing in animals, and it has also been applied to the study of timing in humans. It is, however, very different to the way in which learning has been described earlier in this volume as the acquisition and weakening of associations between events. In the clock based system, there are no associations to speak of; instead the passage of pulses and durations between different memory stores and a decision making process serve to allow the animal to express its learned behaviour. What are we to make of this disparity? One possible answer is that association and timing are entirely different psychological systems, controlled by entirely different processes. One system tells you what goes with what (association), the other tells you when things are going to happen (timing). Although possible, this answer is not very appealing. As we noted at the start of this chapter, it is not much use knowing that food is associated with a light if we don't know when food is going to be delivered—so at some level, the kind of information that is provided by association has to be integrated with timing. Unfortunately, it is not entirely straightforward to see where associations would fit into the clock based model that is shown, for example, in Figure 13.

Perhaps, then, we need to abandon using one system or another, and attempt to explain the principles of learning *and* timing with just one system—based on just a clock *or* just association. Both of these approaches have been attempted. The psychologist Randy Gallistel has taken the first approach. He has argued forcefully against the concept of association, and has proposed, instead, that learned behaviour is a consequence of an animal remembering the duration of events in the environment as well as the rates of occurrence of things like unconditioned stimuli or reinforcement. Although the details of Gallistel's approach to learning are beyond the scope of this book, it does provide, in places, a remarkably parsimonious explanation for conditioning and learning. However, it also has its shortcomings. In particular, it struggles to provide a complete explanation for how animals come to acquire and extinguish conditioned responding. Experiments I have conducted as well as those by other psychologists have shown that animals which have experienced equal rates of reinforcement *and* equal durations of stimuli can still come to acquire and extinguish their conditioned responding differently. This has led a number of psychologists to suggest that learning and timing are underpinned by a system that is based on associative learning.

Timing by association

In order to explain how animals might learn about time by association, we must first consider what we mean by a stimulus.

Is it a discrete event that is unaffected by the passage of time? Almost certainly not—long exposure to a stimulus can change an animal's response to it. For example, if a light is turned on, an animal (a rat *or* a human) might at first be quite startled by it, and perhaps orient their attention towards it. After about a minute, however, this orienting response will diminish, and the light will seem less salient. Thus there has been a change in the animal's response to the stimulus and, potentially, also in its perception of the stimulus. In each case, however, there has been a change in the circumstances of the stimulus with the passage of time—in other words, the start of a stimulus is rather different to a time midway through the stimulus, which itself is rather different to the end of the stimulus, say, a minute later.

For the sake of simplicity, let us assume that an animal's perception of a stimulus changes as it continues to be presented (we could, if we wanted, suggest that it doesn't—instead we could argue that it is just behaviour in reaction to a stimulus that changes—it does not substantially change the argument that is about to be made). Under these circumstances, then, what an animal sees will be different at the end of a long stimulus when it is rewarded with food compared to earlier on during the time when it was not being rewarded. Consider Figure 14, where I have represented how the perception of a stimulus might alter with continued exposure by changing the shading. During the training trials, a relatively short amount of exposure (represented by the mid-grey shading) will be paired, and consequently associated, with food reward. During peak trials, then, responding will be most substantial at the point in the stimulus that has the strongest association with food—the mid-grey zone. The way in which the stimulus is perceived before or after this point will not have been paired with food, and therefore will be less able to elicit responding.

There is, however, a sleight of hand in the previous paragraph. It is not the case in Roberts's peak procedure experiment that the

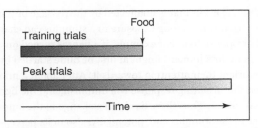

14. An associative model of timing. The perception of a stimulus changes as it is experienced through time.

rats responded in an all or nothing fashion. Responding was graded around about twenty seconds—there was still some responding at times when food had never been delivered to the animal—such as at thirty seconds. Yet the explanation offered in the previous paragraph implies that a response will occur only when the light is perceived in a certain manner. How do we resolve this issue? Different theories of timing make different assumptions, but one way in which this issue can be resolved is to appeal to generalization. Recall that, in Chapter 1, we described an experiment by Guttman and Kalish in which pigeons were rewarded for pecking a key that was illuminated yellow at a wavelength of 580 nm. During subsequent test trials the pigeons pecked frequently even in the presence of colours that were similar to that previously trained, but less so in the presence of colours that were quite different. This result was explained by suggesting that multiple features of a stimulus are conditioned during training, and that these are also present in stimuli that are similar to the training stimulus (permitting generalization), but not present in stimuli that are quite different (preventing generalization). The same analysis can be applied to timing. As a stimulus continues to be exposed, its features may only gradually change, and so intervals that are quite close to the duration that was reinforced during training will be perceived as similar to that duration, thus encouraging the generalization of conditioned responding.

Space

In order to get to work today, I had to drive about 17 miles from my house along (often) congested roads before reaching the outskirts of Nottingham and from there through further urban congestion before reaching the University of Nottingham campus. There is a seemingly never-ending collection of roadworks between my house and the University and, in desperation, I often find myself trying to navigate around traffic jams in order to more quickly reach the blissful calm of my office. This is by no means a simple feat, as anyone who has ever managed to get themselves lost whilst driving without the aid of satellite navigation will appreciate. So how exactly do people learn to navigate around their environment; and furthermore how do animals do it? How do cats, for example, learn to return to the homes of their owners after a long night of debauchery and mouse worrying? Is it really reasonable to suppose that the acquisition of simple associations between stimuli or responses underlie this behaviour too? Maybe; maybe not.

Tolman's short-cut experiment: cognitive maps

An early, and influential, study of spatial navigation was conducted by Edward Tolman and colleagues in 1946. They placed rats into a relatively complex maze at point A on the plan view shown in the left-hand side of Figure 15. In order to find food, the rats had to pass through points B, C, D, E, F, and G. After a number of trials in the maze, the rats were running quickly through it and acquiring food in the goal location. On the basis of associative theories of instrumental conditioning, we might say that the rats learned to find the food through a series of simple stimulus → response associations. Such as 'point C → walk forward', 'point D → turn left', 'point E → turn right', and so on until this chain of associations between stimuli and responses was rewarded by food at the end of the maze. In order to test this possibility Tolman

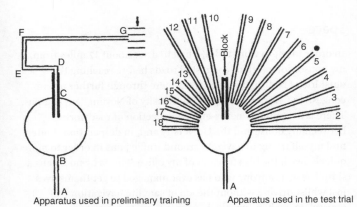

15. **Tolman's maze: the apparatus used by Tolman and his colleagues to investigate cognitive maps in rats.**

modified the maze (see the right-hand side of Figure 15), so that the exit at point C was blocked, and eighteen other exits were provided instead. If the rats had acquired a series of stimulus → response associations, then the rats should perform at chance, selecting one of the eighteen exits at random or, at best perhaps, exits 9 and 10, which were either side of the now blocked exit from training. Tolman's rats did nothing of the sort. Instead, during the test, 36 per cent of the rats left the maze via exit 6 (which, given that there were eighteen possible exits, is a higher percentage than one would expect on the basis of chance).

The results of Tolman's experiment stimulated a number of authors to suggest that rather than using simple associative connections to learn how to navigate to a goal location, animals (perhaps including humans) come to acquire a sort of plan view map whilst navigating that represents the relationships between surfaces and objects in their environment. This idea of animals possessing a plan view representation of their world became known as a spatial or *cognitive map*, and has been very influential in the study of spatial navigation. It is relatively straightforward to see how a cognitive map would aid the transfer of learning from training to test in Tolman's study. By possessing a cognitive map, the rats could calculate the bearing of the goal location from their current position and thus take a short-cut to the goal. It is worth dwelling for a moment longer on some features of Tolman's maze, however, because, upon further inspection, the results of this experiment are not quite as clear as they first appear. During the training stage of Tolman's experiment a stimulus (a light) was located behind the goal location (point G on Figure 15). A similar light was also present in the maze for the test trials, and happened to be located relatively close to the end of exit 6 (it was actually located at the end of exit 5). Upon being at point F during training, Tolman's rats would be rewarded for walking in a straight line along an arm of the maze towards the light. At test then, the rats may have been more likely to leave the maze by exit 6 because this response was very similar to the last response that was reinforced during training (walk along an alley with a light at its end). Tolman anticipated this criticism of the experiment but rejected it, noting that, if anything, the rats should have left the maze via exit 5, not 6, if they were simply conditioned to walking towards the light. This argument has some merit; however, one should bear in mind that Tolman's experiment has not been replicated without the presence of the light in the training and test stages.

Viewpoint-independent navigation

Since Tolman's influential study, rather better evidence for the existence of cognitive maps has emerged from studies of viewpoint-independent navigation. The instrumental conditioning explanation for spatial navigation can be classified as an *egocentric* explanation because it uses stimuli from the viewpoint of the animal in order for it to navigate (e.g. turn right when it perceives a stimulus). In contrast, a cognitive map is *allocentric*, for in this case the animal is navigating using information from a viewpoint other than that which was explicitly experienced during training.

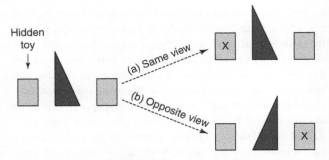

16. Viewpoint independent navigation. Nardini et al.'s viewpoint-independent navigation experiment: 'x' marks the box where the toy was hidden in the test trials.

Marko Nardini and his colleagues have provided striking evidence that human children might rely on the latter when searching for hidden goals. They reported an experiment in which 5-year-old children were placed into a square room and allowed to observe a toy being hidden in one of two boxes that were separated by approximately 60 cm, and between which was an asymmetrical wedge-shaped landmark (see Figure 16). After the children had observed the toy being hidden they were required to retrieve it from one of two starting positions, either (a) the position from which they had observed the toy being hidden or (b) a position on the opposite side of the landmark and boxes from where they originally observed the toy being hidden. In both of these tests, children visited the correct box significantly more often than would be expected on the basis of chance.

It is difficult to construct an explanation for Nardini's experiment that relies on the children using an egocentrically defined stimulus to guide their responses. If they had employed such a strategy then during the test trials in which the children began from a starting point on the opposite side of the landmark and boxes, the children should have chosen the incorrect box rather than the correct box, as this would be the same response as made during training. Instead, Nardini and colleagues suggest that these results could be understood in terms of a more allocentric cognitive map.

Interfering with navigation

In an influential book, the Nobel prize winner, John O'Keefe, and Lynn Nadel further developed the idea of the cognitive map. They suggested that a cognitive map holds information about the position of objects and important places within an environment, but that learning the map is not governed by associative principles. Instead, cognitive maps are continually updated so that if a novel object is added to the environment it will be rapidly represented within the cognitive map and, importantly, *without any interference from other cues that happen to already be present in the map.* This is a curious suggestion, for it implies that if an animal were to first learn that one landmark (e.g. a tree) was a good predictor of where food was hidden then this knowledge should have absolutely no influence on the ability of a new landmark (e.g. a bush) to take control of spatial navigation should it subsequently be added to the environment. In other words, the cognitive map should acquire and represent redundant information—it should be immune to blocking. In one sense this proposal fits in with the way in which maps are made in the real world. A hiker's map, for example, might contain a wealth of information, only some of which they might use to navigate whilst on a country walk—information about the location of a national park boundary is, for most of the time, redundant information to them—yet it is still included on the map. However, the cognitive map is not the same thing as a map used in the real world, it is a psychological construction that is the consequence of exploration in the environment. It is therefore important to ask whether the cognitive map also encodes redundant information—is it immune to blocking? If it was, then it would be set apart from other forms of learning, such as the conditioning experiments conducted by Kamin that were described in

Chapter 3. Furthermore, it would also be at odds with associative theories of learning.

This idea from O'Keefe and Nadel has not stood the test of time particularly well. There are now a number of experiments conducted with a variety of animals (including the ubiquitous laboratory rat as well as humans) that show that spatial navigation is susceptible to blocking. For example, in a blocking experiment conducted by Ken Cheng and Marcia Spetch, honeybees were trained to find sugar water that happened to be placed near to a blue and yellow landmark. For a blocking group, but not a control group, this stage of the experiment was preceded by a stage in which the blue landmark was a signal, on its own, for sugar water. At the end of the experiment a test trial was given to the bees in which the sugar water was removed and only the yellow landmark was present. Bees in the control group spent significantly longer in the region of the environment that had previously contained the sugar water relative to bees in the blocking group. Thus learning about the location of food on the basis of a landmark can be blocked, as is anticipated by, for example, the Rescorla-Wagner model of associative learning. Results such as these have, however, caused some psychologists to redefine what they think a cognitive map might be. Randy Gallistel, for example, has suggested that the cognitive map represents the geometric relations among surfaces within the environment. The idea behind Gallistel's suggestion, here, is that the appearance of landmarks (such as a tree) might change from one season to the next—making it a rather redundant predictor of where resources like food might be located. It makes good sense, then, that landmarks should be susceptible to effects like blocking. He argues, however, that the geometric relations among objects within an environment, or the geometry of the environment itself, remains constant through time, which has raised the possibility that only learning about the *geometry* of an environment should be impervious to blocking. Whether this proposal by Gallistel has any merit is something that is currently undergoing scientific scrutiny. For the time being there is evidence both *for* the proposal (learning about geometry is impervious to blocking), and evidence *against* it (learning about geometry can be blocked). Only further investigation will allow us to understand fully the extent to which the principles of associative learning extend to spatial learning.

Animals and people are quite competent at learning where and when things will be present. This information provides a richer source of information about the world than just association, which, on its own, is not particularly useful—particularly if you want to act upon the information that is provided by associative knowledge. Furthermore, it is now reasonably well established that all of this information can be combined by animals—what is going to happen, where, and when—into something that resembles a single episodic memory of an event. A debate that has permeated both the psychology of timing and spatial navigation is whether or not the acquisition of these forms of information takes place in dedicated systems—a clock and a map—or whether, instead, they can both be understood in terms of a more general associative learning system. The answer to this debate probably won't be resolved any time soon; but if the future of this debate in any way resembles its past, it will be a fascinating scientific enquiry.

Chapter 5
When learning goes wrong

Learning can be very adaptive. It allows organisms to anticipate future events, be they nasty or nice, so that they can be either avoided or approached. Learning permits organisms to make responses in order to acquire desirable things, allowing them to gain control over their environment. Learning informs organisms where things will be, and when they will be available. Learning even helps organisms to link together relatively unimportant events in the world, thus permitting them to bind otherwise disparate elements together into a more structured whole. However, the behaviour of organisms is not always adaptive; sometimes behaviour works against the interests of the organism—even to the point of self-destruction. Take, for example, mental health issues in human beings. Even the most committed geneticist would admit that mental illnesses such as schizophrenia, depression, drug addiction, or anxiety are only partly a consequence of inheritance. There is also some aspect of the way in which individuals interact with the environment that means that they acquire behaviours which are maladaptive or distressing, and which can have a significant impact upon their life, and the lives of their friends and family. Learning, for some reason, sometimes goes wrong; and it can do so in two ways. First, the properties of learning can go awry, so that inappropriate associations develop. One such example occurs in schizophrenia, a disorder that has long been thought to be connected with a deficit in the acquisition of associations. Individuals with schizophrenia appear to form associations between events when non-schizophrenic individuals do not, and it has been suggested that these aberrant associations may form the basis of their psychosis. Alternatively, learning can progress normally—exhibiting the same properties as learning does when it supports adaptive behaviour—but instead it results in the acquisition of behaviour that is maladaptive. There are many examples of this class of maladaptive learning, the most frequently described example (at least in psychology text books) being the acquisition of phobias, in which an ordinarily safe and (essentially) neutral stimulus acquires the capacity to evoke fear as a consequence of being associated with some unpleasant event. For example, a fear of dogs may be acquired as a consequence of once having been bitten or frightened by one. However, in this chapter I will focus on two instances of maladaptive learning which are often given less attention in introductory text books, but which nonetheless can have a dramatic effect on the quality of a person's life. These two instances are (1) the learned side effects of cancer treatment; and (2) drug addiction. I shall describe these two instances of maladaptive learning first before moving on to the role of atypical learning in schizophrenia.

Learned side effects of cancer treatment

According to Cancer Research UK, somebody is diagnosed with cancer every two minutes in the UK. Thus, if, like most people, you have a reading speed of around 200 words per minute then from the start of this chapter to the time you read these words, somebody new will have been diagnosed with cancer in the UK. The cytotoxic drugs that are used in chemotherapy to kill cancer cells have notoriously unpleasant side effects, and two of the most common side effects, affecting around half of the patients undergoing chemotherapy, are nausea and vomiting. We have seen in a number of experiments in the earlier chapters of this book how nausea can very quickly result in the acquisition of an aversion, particularly to foods—even when the interval between the consumption of the food and the induction of nausea is many hours. From this perspective, then, it is not too surprising that patients undergoing chemotherapy often report a loss of appetite: the meals eaten through the normal course of the day could serve as a conditioned stimulus and be the target for an association with the nausea induced by chemotherapy (the unconditioned stimulus).

Loss of appetite in cancer patients has been investigated quite extensively by Irene and Irwin Bernstein and their collaborators. To demonstrate, under more controlled circumstances, that chemotherapy treatment can result in the acquisition of a food aversion, Irene Bernstein and Mary Webster asked adult outpatients at a cancer clinic to eat a distinctively flavoured ice cream between fifteen and sixty minutes before their chemotherapy session. The idea was that the side effects of chemotherapy could potentially serve as an unconditioned stimulus that would support the acquisition of a conditioned response to the ice cream, which in this case was the conditioned stimulus. This is precisely the result that was observed by Bernstein and Webster. After only one pairing of the ice cream with the chemotherapy, patients reported an aversion to the ice cream (the conditioned response) and ate less of it. Other studies conducted by this research group have also showed that this effect is not only restricted to adults, but is also present in children undergoing chemotherapy. Furthermore, an aversion that is acquired to one food may often generalize to other foods that taste similar to it, thus giving the impression that the food aversion is quite general, even though it is the consequence of only one type of food being paired with nausea. Food aversions are also acquired in the full knowledge that the cause of the illness was not in fact the food, but rather the cytotoxic drug used during the chemotherapy. Conditioned food aversions, it seems, are not particularly susceptible to rationalization.

It seems, however, that not all foods are equally likely to be the target of learned aversions. Using questionnaire studies Bernstein and colleagues have reported that foods that are particularly rich in protein, such as meat, poultry, and fish, account for more aversions in the normal population than foods that are high in carbohydrates—presumably because protein-rich flavours are more salient. This has important repercussions for chemotherapy patients as it implies that the nutritional quality of their diet could be affected by their learned flavour aversions. That learned taste aversion might be acquired so quickly and easily—particularly to very salient foods—does however suggest a way in which the effects of learned food aversions could be limited. This is called 'scapegoating'.

The scapegoat effect

A scapegoat is something that is made to bear the blame for others. Ordinarily we think of scapegoats as people who have been unfairly blamed for some misfortune; however, other things can also serve as scapegoats. In the context of the current discussion a scapegoat is some stimulus that is to be used to bear the blame for the nausea induced by chemotherapy. In doing so, if the stimulus is chosen appropriately, then it will reduce the spread of the aversion to other foods in the diet, and consequently reduce patients' reports of a general loss of appetite. We already know that mere instruction about the actual cause of the illness following chemotherapy is unlikely to serve as an effective scapegoat, consequently one approach to establishing a scapegoat is to present patients with a novel and distinctive food prior to chemotherapy sessions. The reasoning behind this procedure is that this food will be particularly salient at the time when nausea is experienced, thus overshadowing the flavour of any other foods that the patients might have normally consumed during the day—protecting these foods from the acquisition of an aversion. This is precisely the effect that Graciela Andresen and colleagues observed. They gave chemotherapy patients either a familiar

	Changes in hedonic ratings		
	Increased	Decreased	No change
Novel group	22	11	20
Familiar group	11	24	22

17. **The scapegoat effect. Consumption of a novel food prior to chemotherapy helps protect other foods in the diet from becoming unpleasant.**

(cookies) or novel (halfa) food ten to fifteen minutes before their chemotherapy treatments. A questionnaire given to the participants before and after this treatment was used to assess how much they liked different kinds of foods in their diets. Figure 17 shows some of the results from their study. The patients who had consumed the novel food prior to chemotherapy actually reported that they liked more of their normal foods after the treatment than they had before—the halfa had served as a scapegoat. In contrast the patients given the familiar food prior to chemotherapy showed the opposite pattern—they showed a reduction in the number of foods that they liked.

Anticipatory nausea and vomiting (ANV)

Although patients undergoing chemotherapy treatment can acquire an aversion to foods that precede this treatment, another aversion that patients can display is to the context of the treatment clinic itself. Mere exposure to the sight, smell, or sound of the clinic before the scheduled chemotherapy session can be enough to cause a feeling of sickness or even induce vomiting. This can be a significant problem—not least because it has an impact on the quality of life of the patient, but also because it can be a factor that influences whether he or she continues with the treatment. Like the flavour aversions described earlier in this chapter, ANV has, at least in part, been attributed to the effect of conditioning—the context of the clinic serves as a conditioned stimulus that is paired with the nausea (the unconditioned stimulus) and potentially the vomiting (an unconditioned response) produced by the cytotoxic drug. Consequently the clinic can come to elicit the conditioned response of ANV prior to the onset of the chemotherapy treatment. A number of variables that are known to affect conditioning also affect the acquisition of ANV. For example, an increase in the incidence of ANV is seen in cases where more severe nausea and vomiting is experienced *after* the treatment. This is consistent with both animal and human studies that have shown that conditioning is more successful with a more intense unconditioned stimulus. Furthermore a longer infusion of the drug is also associated with a greater incidence of ANV. This effect can be attributed to there being a longer pairing of the conditioned stimulus (the clinic) with the unconditioned stimulus (the side effects of the drug).

One of the best ways of preventing the acquisition of associative learning from occurring in the first place is to ensure that one of the items of the association is not present. Thus, in the absence of an unconditioned stimulus, there is no reason for an initially neutral stimulus (in this case the hospital clinic) to become a conditioned stimulus. Consequently, if chemotherapy treatment can be given in the absence of its emetic side effects, then ANV should be reduced. Early studies of ANV suggested that this might not be the case—ANV was observed in a number of cases even when patients were given drugs to reduce nausea and vomiting. However, more recent studies, particularly those that have used more effective anti-emetic drugs, have been more positive. For example, Aapro and colleagues reviewed the cases of 574 patients who had received the anti-emetic granisetron during their chemotherapy cycles. Fewer than 10 per cent of the patients displayed symptoms of anticipatory nausea and fewer than 2 per cent had symptoms of anticipatory vomiting. These figures suggest, thankfully, that the incidence of ANV is on the decline, as earlier studies had reported ANV rates as high as 33 per cent.

The acquisition of drug addiction

In Chapter 1, two categories of classical conditioning were described: (1) aversive conditioning—where the unconditioned stimulus was something unpleasant; and (2) appetitive conditioning—where the unconditioned stimulus was something nice. The learned side effects of cancer treatment constitute an example of maladaptive learning that clearly fits into the first of these two categories—nausea and vomiting being things that are undesirable. However, maladaptive behaviour can also be acquired as a consequence of learning about something that is desired—and some of the most desirable things that people seek out are 'drugs of addiction', such as the so-called hard drugs like cocaine and heroin, or the more commonly available drugs, such as alcohol or tobacco. Drugs such as these will often support a number of behaviours that sustain an addiction. They produce *cravings* (in which the user reports, or experiences, an overwhelming drive to take the drug) and also a *tolerance* (in which the user requires more of the drug in order to achieve the same effects as they did when they first took it). In this section of the chapter, I shall consider the acquisition of these two properties of drug user behaviour, describing the role of conditioning in drug seeking, before considering how extinction might alleviate some of these problems.

Classical conditioning of cravings and tolerance

Drug users report that a variety of stimuli in the environment can elicit cravings, and these are frequently the sorts of paraphernalia that are associated with the drug itself, such as the sight of the drug, a bank note, or even the drug dealer themselves. That such objectively neutral stimuli can evoke such a strong craving response in drug addicts is consistent with the idea that these stimuli may control behaviour through classical conditioning. Stimuli such as the sight of a dealer (the conditioned stimulus) will be paired with the unconditioned stimulus of the drug (e.g. cocaine). With repeated pairings of these two stimuli, they will become associated and the initially neutral stimulus will come to evoke a substantial conditioned response. An experiment reported by Richard Foltin and Margaret Haney showed that a variety of conditioned responses could come to be acquired to initially neutral stimuli that were paired with cocaine in habitual users. In this experiment, a neutral stimulus (e.g. the odour of peppermint) was paired with either cocaine or a placebo. After two weeks of this training, the stimulus paired with the drug was able to evoke various physiological responses—including a change in heart rate and skin temperature, as well as more subjective, self-reported measures of craving (e.g. the user's reports of 'wanting' cocaine). Although these results are compelling, and are consistent with the conditioning analysis of drug addiction, they do not tell us whether a similar effect would also be evident in people who do not have a drug habit. This is an important point because the classical conditioning explanation for drug use is a model of how the addiction is acquired—not just how it is sustained. Fortunately, a study similar to that conducted by Foltin and Haney was described by Steven Glautier, Colin Drummond, and Bob Remington, and this showed that conditioned responses of changes in skin conductance and heart rate could be acquired to the colour of a liquid that was paired with the consumption of alcohol, as well as more subjective measures such as participant's arousal. Importantly, however, this experiment was performed with individuals who had no reported history of drinking problems. Together, then, these two studies show that stimuli that, at first, have little ability to elicit a response can come to control substantial physiological reactions as a consequence of being paired with a drug. These experiments are therefore consistent with a conditioning explanation for the acquisition (and maintenance) of drug cravings.

Many drugs of abuse come to be associated with the development of a tolerance to the drug. Thus, an individual who is experienced with opiates, such as heroin, can survive a dose which would kill a drug-inexperienced person. One explanation that has been proposed for the acquisition of drug tolerance comes from our understanding of some of the effects of classical conditioning. Recall that, in Chapter 3, we discussed how the representation (or perception) of stimuli can change as a consequence of learning. In particular it was noted that an expected event serves as a less effective unconditioned stimulus than an unexpected event. The unconditioned responses to a variety of stimuli—from puffs of air to the eye to the sight of a light—are diminished when they are expected. We have just seen in the experiments in the previous paragraph that environmental stimuli can come to be established as predictors of drugs of abuse. This then leads to the possibility that the drugs themselves may become less effective when they are anticipated—that is to say, in the presence of the sorts of stimuli that come with drug administration, a conditioned tolerance should occur.

Shepard Siegel has provided some of the most compelling evidence for this proposal. In an experiment conducted by Siegel and his colleagues, rats were given injections of heroin in one room (room A) and injections of dextrose in another room (room B). A control group was given dextrose in both rooms. The idea behind this experiment was that room A should become associated with heroin, but room B would not. Consequently, when placed in room A, the rats would anticipate heroin and its effects should be diminished. To test this prediction, Siegel and his colleagues gave all the rats an overdose level of heroin in either room A or room B (see Figure 18). We can see that the dose given to the rats was an effective overdose as almost all of the rats in the control group—who had never experienced heroin before—died when given it. The mortality rate in the drug-experienced rats was much lower. Most interesting, however, was the mortality rate of the rats given the overdose in room A or

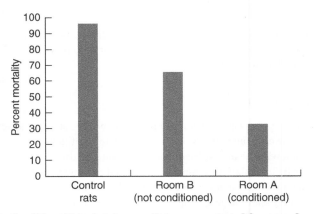

18. **Conditioned heroin tolerance. Rats were protected from overdose by being administered the drug in a context in which they were familiar with it.**

room B. When the overdose was given in room B—where heroin had never been experienced before—then the mortality rate was around 65 per cent; however, when the overdose was given in the room where the drug had previously been taken (room A), the mortality rate was much lower—around 32 per cent. These final two groups had had identical experience with heroin—the only difference between them was whether they were receiving an overdose level of the drug in the presence of the same stimuli as during their earlier training. Evidently, being in the presence of cues that had been paired with heroin in the past diminished the effectiveness of the drug. A conditioned tolerance had been acquired.

One might question, however, why the rats given the overdose level of heroin in room B showed a lower mortality rate than the control rats. Room B had never been paired with heroin prior to the overdose, and yet the effects of heroin seem to be diminished in this group. Why? One possible answer to this question is that there may be an effect of exposure to heroin that is not based upon conditioning, which provokes the development of a tolerance. Alternatively, it may be that part of the conditioned tolerance acquired to room A generalized (see Chapter 2) to room B based on the shared features of the two rooms.

Drug seeking

As powerful as classical conditioning may be for making a drug user crave a drug and develop a tolerance for it, it does not explain why the user subsequently seeks out the drug and then takes it. To explain this behaviour we need a form of learning in which the user can acquire a response that results in them obtaining their desired drug. For this reason, instrumental conditioning has been argued to form the basis of *drug-seeking behaviour*. The logic behind this application of conditioning is that the response that results in the administration of the drug results in the establishment of an association between it, environmental stimuli, and the reinforcing properties of the drug. You may recall from our description of the experiment by Christopher Adams in Chapter 2 that the association between the response and the reinforcer could be revealed in instrumental conditioning by devaluing the reinforcer once instrumental conditioning had been established. Lee Hogarth and Henry Chase used a variation on this procedure in an experiment to examine whether instrumental behaviour in human smokers is similarly controlled. Participants were first trained to press one of two different buttons for either chocolate or cigarettes. Following this training, participants were given the opportunity to smoke freely until they felt as if they didn't want to smoke any more—the idea being that the value of the reinforcer would be reduced following this procedure. Finally, the smokers were allowed to press the two keys, but without the reward. The results showed that all participants reduced their choice of the tobacco key following the devaluation procedure. It seems that for cigarette smokers, responding that got them a cigarette—that is to say, their drug-seeking behaviour—is under the control of an association between a response and a reinforcer.

The results of Hogarth and Chase's experiment might come as something of a surprise, for we tend to think of drug seeking as a habit, not something that is particularly mindful or executed with knowledge of the goal of the response. It is also surprising when considered in the light of the experiment by Adams which revealed that not all instrumental behaviour was a consequence of associations between the response and the reinforcer. Recall that he showed that with substantial exposure to instrumental conditioning, behaviour became insensitive to reinforcer devaluation—thus habits in instrumental conditioning *can* be acquired. Whether a similar transition from goal-directed behaviour to habit occurs in drug addiction remains to be determined in human addicts, but for animals, the answer seems to be 'yes'. In an experiment by Andrew Nelson and Simon Killcross, rats were given exposure to amphetamine every day for a week. Following this treatment, the rats were given standard instrumental conditioning training in which they were rewarded with sucrose for pressing a lever. Subsequently, sucrose was devalued by pairing its consumption with illness. In a final test, in which the rats were given the opportunity to lever press (in the absence of sucrose), Nelson and Killcross found that the amphetamine rats showed no sensitivity to reinforcer devaluation. However, control rats—those that had had no exposure to amphetamine—did show sensitivity to reinforcer devaluation. This is an interesting result. It shows that exposure to a drug, in this case amphetamine, can substantially alter the way in which organisms subsequently learn to acquire things in the world. Following chronic amphetamine exposure, animals are less able to appreciate the consequences of their actions. This result therefore goes some way to understanding two features of addiction: why addictions are maintained even when a tolerance has been acquired to the drug (reducing its value); and why addiction results in such irrational behaviour.

Treatment of addiction

As we saw in our description of classical conditioning in Chapter 1, the conditioned responses acquired to a stimulus can be weakened—extinguished—by presenting the conditioned stimulus in the absence of the unconditioned stimulus. This forms the basis of *cue-exposure therapy*, which is used to treat a number of instances of acquired maladaptive behaviour. In the case of substance abuse, the drug addict is exposed to drug-associated stimuli, but in the absence of the drug itself, in order for the conditioned responses elicited by the drug-associated stimuli to extinguish and the cravings to subside. Many studies have examined whether cue-exposure therapy is effective in treating drug addiction. Some of these studies suggest that it is an effective treatment; others, however, have suggested it has no effect at all. Cynthia Conklin and Stephen Tiffany reviewed these studies in a meta-analysis of cue-exposure experiments. Disappointingly, they found that the overall size of the effect of the therapy was not significantly different from chance.

On the one hand, it is surprising that cue-exposure therapy is so ineffective in treating drug addiction given the effectiveness of extinction in studies of animal (and indeed human) conditioning. On the other hand, however, the extinction of conditioned responding is a fragile phenomenon: merely conducting extinction in a context that is different to that in which conditioned responding was acquired is sufficient to disrupt its permanence. Cue-exposure therapy is rarely conducted in the same context in which the original drug addiction was acquired. Therefore, even though drug addicts may be seemingly clear of their habit following cue-exposure therapy, should they be exposed to drug-associated stimuli in the context in which they originally acquired their habit, then our theories of extinction predict that there is a high chance of *relapse*, and this unfortunately frequently occurs.

Disrupted learning in schizophrenia

In the previous two sections of this chapter we have seen how learning can serve to produce maladaptive behaviour. In the cases of the learned side effects of cancer treatment and in drug addiction, this maladaptive behaviour was a consequence of learning working in a perfectly normal manner. That is to say, learned behaviour was acquired, and sustained, in a manner that was comparable to the way in which learning is acquired and sustained during more adaptive circumstances. In the final section of this chapter, however, I want to consider a case of learning going wrong in an abnormal manner—in this case, in schizophrenia.

Schizophrenia is the name given to a collection of psychiatric symptoms, which include (but are not always limited to) delusions, hallucinations, and muddled thoughts; social withdrawal; and anhedonia (an inability to feel pleasure in normally pleasurable things). It occurs in approximately 1 per cent of the population, with men and women affected equally, and with the most common age of diagnosis being between 15 and 35 years. The name 'schizophrenia' was coined by the Swiss psychiatrist, Eugen Bleuler, in 1911 and has Greek roots, with 'schiz' meaning split and 'phrene' meaning mind. It is, then, not entirely surprising that schizophrenia is often incorrectly thought to mean 'split personality'. This would however be a misnomer; instead it is better to regard the name as describing the fragmentation of thinking that people with the disorder suffer. In particular, Bleuler noted that one of the fundamental components of schizophrenia was a loosening of associations, which he described as:

> Contradictory, competing, and more or less irrelevant responses [that] can no longer be excluded. (Bleuler, *Dementia Praecox, or the Group of Schizophrenias*, p. 511)

This is an interesting observation for, as we saw in Chapter 3, one of the things that associative learning seems to provide human and non-human animals with is the ability to tighten up their associations so that they learn more about features of the environment that are relevant predictors of outcomes, and to tune out redundancy. Take, for example, Kamin's blocking experiment that was described in Chapter 3. This is an example of stimuli competing for an association, as learning is prevented with one stimulus when it is accompanied by another stimulus that is a better predictor of a subsequent event (in Kamin's case, a shock). It would seem to follow from Bleuler's early description of the disruption that is observed in schizophrenia, then, that individuals with this condition should be less able to learn to ignore irrelevant or redundant information. This prediction will be the focus of this final section of the chapter.

Schizophrenia, blocking, and learned irrelevance

A number of studies, particularly with patients who are in the early, acute, phase of schizophrenia have revealed that blocking is indeed disrupted in these individuals. A particularly clear demonstration of this was reported by Jones and his colleagues in 1997. They recruited people with schizophrenia and control participants without the condition, and gave both of the groups a task that was intended to mimic Kamin's blocking procedure—but without the complication of having to use biologically significant outcomes as the unconditioned stimulus (which might be particularly upsetting for patients). In this experiment, the participants had to rate how likely a film would be a box office hit based upon the name of the actors that were in it. In the first stage of the experiment one actor (let's call her Libby) would be paired with a 'box office hit' then in stage 2 of the experiment Libby would be in a film with another actor (say, Ellie) and this film is also a success. The question of interest is how much do participants rate the box office potential of Ellie? Based on what we know about blocking, it should be the case that participants'

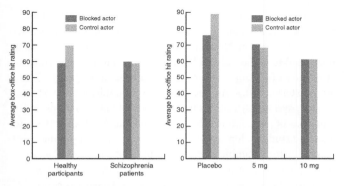

19. **Blocking is disrupted in patients with schizophrenia and also healthy individuals given amphetamine.**

prior knowledge about Libby being a box office hit should prevent participants learning about Ellie when she and Libby are in the same film—blocking. In a control condition, two other film stars were described as being in the same hit film (say, Mark and Siân) but neither of these was used in the training at the first stage of the experiment. Consequently, both of these actors should be learned about. The results of the experiment are shown in the left-hand panel of Figure 19.

Healthy individuals demonstrated a small, but reliable, blocking effect—they learned more about the control actor than the blocked actor. However, participants with schizophrenia showed no blocking at all—they learned just as much about the blocked actor as about the control actor. Although these results are quite persuasive, one might question how direct this effect is. People with schizophrenia differ from healthy participants in a number of ways—not least in terms of the number of medications that they might be taking. Therefore, how can we be sure that it is not the medications that are disrupting blocking rather than the disorder itself? One way in which to address this question is to actually manipulate the presence and absence of schizophrenia itself within an experiment; we can then be more confident that the factor that we have manipulated had an effect on the measure that we are interested in. Now, of course, it is difficult (not to mention unethical) to induce schizophrenia in people; however, certain symptoms of schizophrenia can be mimicked in people by administering drugs. Amphetamine, for example, is known to induce psychotic symptoms including visual hallucinations and delusions of persecution. The right-hand panel of Figure 19 shows a second study by Jones and his colleagues which shows the effects of administering amphetamine (or a placebo control) in healthy participants on blocking. It is hopefully clear that blocking was present in the placebo condition—participants learned more about the control actor than they did about the blocked actor; however this effect was abolished under both doses of amphetamine.

Schizophrenia is known to have a disruptive effect on learning that is broader than just blocking. Studies by Richard Morris, Oren Griffiths, Mike Le Pelley, and Thomas Weickert have also shown that individuals with schizophrenia are less able to learn to ignore explicitly irrelevant information. They presented pairs of seeds on the screen of a computer to participants, who had to predict which of two different types of tree would grow when the seeds were planted. Importantly, half of the seeds were predictive of the type of tree that would grow, whereas the other half were irrelevant—half the time these seeds were associated with one type of tree, on the remaining trials they were associated with the other. Although both schizophrenia and control participants were able to learn this task, an interesting difference emerged in a second stage of the experiment when the seeds were used in a new task. In this new task all the seeds were now equally relevant for predicting which type of tree would grow—the question of interest was whether participants would now have a bias to learning more about the seeds that they had been trained to consider relevant. That is to say, would they be able to tune out the previously irrelevant information and focus on the important detail? For healthy control participants the answer to this question was, 'yes'—they learned more about the previously predictive seeds than the previously irrelevant seeds—even though, in this new task, all the seeds were equally relevant to solving the task. Individuals with schizophrenia, however, showed a different pattern of results: individuals who displayed particularly high levels of psychosis showed no such bias in learning—they learned just as much about the relevant as the irrelevant seeds. A closer look at the data suggests that this effect was a consequence of the schizophrenic participants paying undue attention to the irrelevant seeds.

These studies of blocking and learned irrelevance provide an interesting insight into some of the difficulties that people with schizophrenia must face every day. They suggest that people with schizophrenia are disrupted in their ability to ignore irrelevant information. Without the ability to learn which information is relevant to solving tasks, and to tune out irrelevance, the world must be a particularly confusing place. Consider, for just a moment, how difficult it would be to learn to do even the most straightforward of tasks, such as crossing the road, if your attention was just as captured by the noise of a passing aircraft or the sight of surrounding houses as it was by the movement of cars. Interestingly, these sorts of effects can also be observed in certain parts of the normal population. If one thinks of schizophrenia as the extreme end of a continuum of a personality type then it follows that there may be some individuals in the population who are not symptomatic with schizophrenia, but who nonetheless have personality characteristics that are consistent with some of the traits of schizophrenia. This continuum can, and has, been measured with questionnaires, and a number of studies have now revealed that individuals who are high on this so-called continuum of 'schizotypy' are, like schizophrenia patients, also disrupted in their ability to display blocking and learned irrelevance.

To investigate the reasons why learning and behaviour goes wrong is, perhaps, one of the most interesting questions that can be asked in psychology, and this chapter has highlighted three such instances. We have seen how learning can work entirely normally but, unfortunately, can be applied in such a way as to make it maladaptive—in the cases of some of the psychological side effects of cancer treatment, and the acquisition of substance addictions. We have also examined how learning can be disrupted in the case of schizophrenia, and how this may contribute to learning about inappropriate things. Other cases of mental health disruption are also studied from the perspective of learning and conditioning—including depression, anxiety, eating disorders, and even psychopathy. The World Health Organisation estimates that one in four people in the world will be affected by mental or neurological disorders at some point in their lives. On this basis it is likely that the clinical application of learning will be a particularly useful focus for future research.

Chapter 6
Learning from others

Up to now, we have been considering the way in which learning is acquired in individual animals. However, many species, including humans, are quite sociable and live in groups of one size or another. It is therefore important to incorporate into our understanding of learning whether and how information might be acquired from another organism. In other words, how social learning might take place, and what sorts of factors might influence it.

Adding a second animal into the picture when discussing learning adds some complications, because not only will the researcher have to deal with potential variability in the behaviour of the *recipient* of the learning but also any variability in the behaviour of the animal that is being *learned from*. More fundamentally, however, there is the complication of how exactly to refer to each animal in the various studies on social learning. It is useful, therefore, to introduce some terminology to ease this problem. Two terms that are frequently used when discussing social learning are 'observer' and 'demonstrator'. The *observer* is the organism that is the recipient of the information in the social learning situation. Typically, therefore, this animal is the focus of behavioural measurements that may be used to detect learning. The *demonstrator*, on the other hand, is the animal that is (intentionally or otherwise) providing information to the observer.

These two roles may not necessarily be fixed—social interactions typically result in animals taking on either, or both, roles at any one time. Regardless of this, however, these terms are useful in order to ease the understanding of how learning moves between one organism and another.

Learning what to eat from others

We have seen in earlier chapters how learning can have a significant influence on an animal's diet. Although they are typically cautious about novel foods, animals will, with experience, tend to accept new foods into their diet. However, from what we know about learned flavour aversions, it also clear that should a food be a cause of illness then animals will quickly change their behaviour and learn to avoid the food. Although these behaviours seem, at first sight, entirely sensible strategies for dealing with potential toxins in the environment, a moment's reflection will reveal that learning in this non-social manner may also be detrimental. In order to learn which foods are safe and provide nutrients, and which foods are harmful, the animal has to place itself into a situation that is potentially disastrous. Eating a very toxic plant, for example, provides the animal with no opportunity to later express its learning about the relationship between the taste of the food and its ultimately lethal consequences—as the animal will be dead. What use is learning under these circumstances? The benefit of being able to observe the consequences of food selection in other animals is therefore substantial.

Bennett Galef and his colleagues have provided several demonstrations of an animal's food choice being influenced by the presence of another animal. McQuoid and Galef, for example, showed that observer Burmese jungle fowl preferentially ate out of a bowl that they had previously observed a demonstrator fowl eating from. This effect is not limited to birds or visual observation either. In a further study, Galef permitted a demonstrator rat to eat a distinctive and novel food (e.g. cinnamon). Observer rats were then given the opportunity to interact with the demonstrator rat for half an hour. In a final test, conducted in the absence of the demonstrator rat, the observers were given a choice between consuming cocoa or cinnamon in the absence of the demonstrator. The observers selected the food that had been consumed by the demonstrator. Why might animals choose the food that they have previously seen a demonstrator eating or can smell upon them? One possible explanation for these results appeals to a simple process of habituation, which was described in Chapter 1. The initial response to seeing a novel food may be to avoid it, however, the presence of the demonstrator may draw the attention of the observer towards the food (or the object from which it was eaten). This may then have the consequence of encouraging the observer animal to interact with it later, speeding the rate at which the observer animal's avoidance response to the food is habituated. Although the simplicity of this explanation is compelling, it is unlikely to provide a complete explanation for all instances of socially learned food choice in animals. Further experiments described by Galef have shown that the preference for cinnamon demonstrated during the choice test by the observer rats was not as substantial when the cinnamon had been dusted over the rear end of the demonstrator rat compared to when it had been dusted over its mouth instead. In both of these cases the food was experienced by the observer in conjunction with the demonstrator—and yet the effects of this interaction were very different. We therefore seem to have to appeal to an additional process—beyond habituation—if we are to explain how social learning of diet choice emerges in animals.

Is social learning a form of conditioning?

The fact that many animals have a preference for being in groups suggests that there may be something intrinsically rewarding about the presence of another animal. This reward value may be something that is unconditioned (i.e. not learned), or it may be something that has been acquired by being paired with some other pleasant event (e.g. food or sex). In either case, if we consider the presence of another animal as something akin to a reinforcer in instrumental conditioning, or an unconditioned stimulus in classical conditioning, then it may be possible to develop an explanation for social learning that is based upon the principles of these types of learning. Bennett Galef and Paula Durlach, for example, have suggested that the socially acquired preference for cinnamon described in the experiments in the previous paragraph could be thought of as a conditioned response. Here the food itself (or the food cup, in the case of the experiment by McQuoid and Galef) was the conditioned stimulus that was paired with an unconditioned stimulus of some aspect of the demonstrator's behaviour.

Studies of the social acquisition of fear in monkeys by Michael Cook and Susan Mineka have also placed an emphasis on the role of classical conditioning. They showed that laboratory reared observer rhesus monkeys would acquire a fear response to snakes (e.g. alarm calls, agitation) if they were permitted simply to observe a demonstrator monkey's fear response whilst in the presence of the snake. Mineka and Cook suggested that the snake served as a conditioned stimulus which was associated with the unconditioned stimulus of fear in the demonstrator monkey, and referred this form of learning as *observational conditioning*. Interestingly, not all stimuli seem to be equally able to serve as a conditioned stimulus for observational conditioning. In another experiment, observer monkeys were given twelve sessions in which they were shown a video of demonstrator monkeys expressing a fear response in the presence of either a snake or some flowers. Both before and after this training, the observer monkeys were presented with videos of the snake or flowers (in the absence of the demonstrators) whilst making a food-reach response—the logic being that the observer monkeys would be slower, or less inclined, to search for food if they were afraid. The results of this experiment are shown in Figure 20. Before the

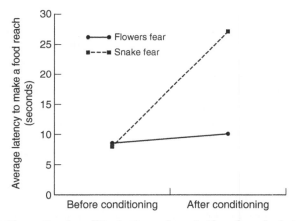

20. Observational conditioning in monkeys. A video of a snake, but not flowers can be associated with fear in another animal.

videos of the flowers and the snake had been paired with the fear response in the demonstrator monkey, the observer monkeys were happy to make a reaching response in order to obtain food. After this training, however, they were much more hesitant—in particular, during the video of the snake. There was very little evidence for the acquisition of social conditioning when the video of the flowers had been paired with observation of a fear response. A very similar pattern of data was observed when direct measurements of fear responses in the observer monkey were taken.

These results are interesting for in both the case of the snake fear and the flower fear conditions of the experiment, the observer monkeys were given trials in which an initially neutral stimulus was paired with the observation of fear in a demonstrator animal; yet only one of these stimuli—the snake—came to elicit fear in the observer monkeys. This suggests that not all stimuli are created equally for learning. In fact, this is quite a well established idea in conditioning. Martin Seligman, for example, suggested that humans have an inherited predisposition (which he called 'preparedness') to rapidly acquire fears to certain objects that may have posed a threat to our ancestors, and that this preparedness might explain why certain fears and phobias are so prevalent (e.g. fear of spiders) when in fact people rarely encounter such things in conjunction with an aversive event. Similarly, it is particularly easy to condition an aversion to flavours with illness, but stimuli such as lights and tones are less successful—again suggesting that some stimuli are more prepared for an association with a particular outcome than others. That a similar effect of preparedness can be observed in social learning, here in Cook and Mineka's study, therefore encourages the idea that this instance of social leaning may be underpinned by a similar mechanism to that which supports non-social learning, such as classical conditioning.

Imitation

In the examples of observational conditioning and diet choice described in the earlier paragraphs, the response that the animals acquired through observation was one that was already in their behavioural repertoire—eating or eliciting fear responses, for example. However, it does not seem beyond the realm of possibility that an animal could also acquire an entirely novel response by observation as well. *Imitation* is a form of social learning in which an animal acquires a new response as a consequence of observation of another animal. For example, Galef and his colleagues allowed observer budgerigars to watch a demonstrator be rewarded with food by making one of two responses: lifting a flap either (a) with its feet; or (b) with its beak. Observer birds that had watched the demonstrator bird lift the flap with its feet performed this response more frequently, using their feet rather than their beak; and vice versa for the birds that had watched the demonstrator access food with its beak. It is difficult to explain imitation in terms of classical conditioning because a new form of response is acquired in this kind of social learning. In classical conditioning the unconditioned response—which the animal is already able to make—comes to be acquired by another stimulus (the conditioned stimulus). With imitation, however, the response that is acquired is novel. Therefore, it has been argued by some psychologists, such as Cecilia Heyes, that imitation is a consequence of instrumental conditioning, which, in the non-social domain, also permits the acquisition of a novel response.

An interesting example of imitation comes from a study by Deborah Custance, Andrew Whiten, and Kim Bard. The first stage of their study lasted three and a half months and involved rewarding two chimpanzees for imitating fifteen different actions performed by the experimenters whenever they said, 'Do this!'. For example, the experimenter would say, 'Do this!' and raise both arms, slap the floor, or touch his or her armpit. If the chimpanzees performed the same action they would be rewarded. Following this training, the experimenters began making untrained actions after saying, 'Do this', such as touching the back of the head, puffing out the cheeks, or interlinking their fingers. The results of the study indicated that on a significant number of occasions, the chimpanzees imitated these, untrained, actions. These results constitute a particularly striking demonstration of imitation.

Sometimes imitation can be quite an automatic thing. Humans, for example, will tend to elicit the same action as another person, even if it happens to interfere with an ongoing task. This effect can be demonstrated on a friend—ask them to open their hand as fast as they can when they see your hand moving. Your friend will be pretty good at doing this so long as your hand happens to be *opening*. However, if you suddenly *close* your hand, your friend will find it much more difficult to open their hand—they will have an automatic urge to do as you have done, despite their requested task being a relatively simple one. An experiment that I conducted with Rosetta Mui, John Pearce, and Cecilia Heyes demonstrated that automatic imitation could be observed in non-human

21. A demonstrator budgerigar stepping and pecking. Black and white stills from the videos that budgerigars watched of a demonstrator either pecking or stepping on a button.

animals as well as humans. In this experiment budgerigars were placed into a testing chamber that had a button located on the floor at one end, and a computer monitor at the other. On this monitor, short videos were played of a demonstrator budgerigar sometimes stepping on the button and other times pecking at the button (see Figure 21). For one group of observer budgerigars—the compatible group—responses that were *the same* as those shown on the monitors were rewarded. So they had to step when they saw a step or peck when they saw a peck in order to get food. For another group—the incompatible group—responses that were different to those displayed on the monitor were rewarded. So these birds had to step when they saw a peck and peck when they saw a step in order to get food.

The results of this experiment were very clear: over ten days of training, birds in the compatible group had little difficulty in solving this problem, making significantly more correct responses than incorrect responses. Birds in the incompatible group, however, had much greater difficulty learning their task—for much of the time they made as many incorrect responses as correct ones. This experiment suggests that when seeing and doing are incongruent, there is substantial interference in the task—which is precisely the automatic imitation effect that is observed in humans.

Mirror neurons

One of the most influential discoveries in recent years in the field of social learning has been the identification of a network of neurons that have a property that makes them particularly suited to being the potential neural basis of social learning. These neurons have been termed *mirror neurons*. These cells were first discovered by the neuroscientist Giacomo Rizzolatti and colleagues in 1992 while they were investigating how neurons fired in the pre-motor region of the brain of macaques when they made movements with their hand and mouth. Quite unexpectedly, they noticed that some of these neurons were activated not only when the monkey was making a hand movement but also when it observed the *experimenter* making a hand movement. Thus, here we have neurons that fire both during the observation of a behaviour being performed by a demonstrator and during performance of the same action by the observer. Since this original discovery, neurons with similar properties have also been observed in other brain regions, such as the parietal cortex, as well as in the human brain. Mirror neurons seem to provide, at the very least, a way in which the brain can encode what is seen and what is done in a common manner—which is presumably a requirement of any explanation for social learning. As Cecilia Heyes wrote, with typical eloquence, mirror neurons

> bridge the gap between one agent and another; to represent 'my action' and 'your action' in the same way. (Heyes, *Where do mirror neurons come from?*, p. 575)

We can contrast two explanations for how mirror neurons came about, one of which emphasizes evolution, while the other emphasizes learning—in particular, associative learning. According to the first explanation, mirror neurons have an adaptive purpose—they exist in order to permit organisms to understand the actions of other organisms. Thus, evolution favoured the natural selection of these neurons as they provided an advantage to organisms in social groups. This view of mirror neurons could be likened to the unconditioned stimulus and unconditioned response in classical conditioning, as no learning is required for mirror neurons to behave in the manner in which they do. According to the second explanation, which has been advanced by Heyes, mirror neurons are an example of associative learning. Consider the case in which you observe yourself making a fist with your hand. Under these circumstances you are receiving input from the visual representation of this action (i.e. what you see) as well as activation of a motor representation of this action (i.e. what you are doing). It just so happens that when you observe yourself performing a response like this then what you see (the stimulus) and what you do (the response) are paired together in time—which, as we have seen, is an excellent circumstance for an association to be acquired between these two events.

One way in which stimuli and responses might co-occur to form a mirror neuron has already been mentioned—by observing one's own responses. Human infants, in particular, seem to spend an inordinate amount of time watching their own hands move—which provides them with excellent experience of a stimulus and response co-occurring. In addition, adult humans will frequently mimic the facial movements of their own children, thus providing the child with further opportunities to experience co-activation of the sight of a response with their own performance of a similar response.

How might the associative theory of mirror neurons be tested? One way in which the merit of the theory could be assessed is to see whether the simple pairing of any arbitrary stimulus with a response made by participants is enough to permit a link to be formed between these events. Parts of the human brain that are thought to form part of the mirror neuron 'network' (e.g. the pre-motor cortex) should then be activated by both the performance of the response and the arbitrary stimulus. This was precisely what was observed by Clare Press and her colleagues in an experiment in which human participants first received training in which they had to make distinctive hand gestures whilst they looked at different geometric shapes (e.g. point a finger whilst looking at a circle or make a fist whilst looking at a triangle). Once participants had mastered this training they were placed into an MRI scanner and the activity of neurons in the pre-motor cortex was measured. Press and colleagues found evidence that the same neurons that were activated by the different hand gestures were also activated by their associated geometric figures. These results are not easy to explain by the adaptation theory of mirror neurons—it is difficult to see why geometric shapes, for example, should assist in the understanding of actions of other animals and therefore be particularly prone to activating the part of the human brain containing mirror neurons. In contrast, these results are entirely in keeping with the associative explanation for mirror neurons in which, in its simplest form, it does not matter what the stimuli and the responses are—only that they should both be activated at about the same time.

One of the classic ways to determine whether something is learned or whether something is innate is to examine whether it is present in very young children. According to the associative explanation of the origin of mirror neurons, imitation is something that must be learned. Therefore there should be little or no evidence of imitation in very young children—as they have not yet had an opportunity to form associations between their responses and the sight of those responses being performed. In contrast, according to the adaptation explanation imitation should be present from birth. Unfortunately, the evidence relating to this issue is not particularly clear. Very young human children will only reliably imitate one action—tongue protrusion. However, it remains to be determined whether this effect actually is a consequence of imitation or whether it happens to be a more general response that the baby is making when it is aroused.

Like classical and instrumental conditioning, social learning is a change in behaviour as a consequence of experience. It therefore fits in with the definition of learning that was provided in Chapter 1. However, the nature of this learning is rather different—it is experience of observing another organism. Social learning has obvious adaptive value—especially when it comes to diet choice. Furthermore, some researchers have suggested that social learning may be underpinned by the same forms of conditioning-like processes as non-social learning. However, social learning continues to be a difficult form of learning to fully understand, not least because the actions that we might observe in another individual may not match the way in which we perceive those actions when performed by ourselves. Associative theories of social learning can address some of these issues, but it continues to be a field of psychology that stimulates much research.

Chapter 7
Surely there is more to learning than that?

At the start of this book, learning was defined as a relatively permanent change in behaviour as a consequence of experience. The subsequent chapters in this book have provided a variety of examples of learned behaviour, some of which is pleasant, some of which is scary; some helpful, and some downright maladaptive. Learning has been described in cases when stimuli are present, as well as when they are absent. Learning has been described in the context of space, the context of time, and in terms of the social environment. At face value, these various types of learning are very different. Compare, for example, the acquisition of morphine tolerance in rats to automatic imitation in budgerigars—quite distinctive behaviours. However, time and again we have seen how some psychologists have made an attempt to explain different forms of learning with a common principle—association. This principle explains learning with a simple idea: changes in behaviour are a consequence of the formation or weakening of links between things. The items of association may be many and varied—they may be a tone and a shock, a lever press and food, or the sight and the performance of a yawn. This variety of associable events allows associative theories to provide an explanation for the many apparently different guises that learning can take. At the heart of this principle, however, is a very simple psychological building-block: an association, something that is no more complicated than a link in a chain.

A moment's reflection on the types of thought processes that you have probably gone through whilst reading this book might, however, make you conclude that it is too simplistic: association surely can't be all that there is to learning. You might think that on a number of occasions you have made some very deliberate acts of decision making—such as a conscious decision to keep on reading. You might also reflect back to a time when you have learnt the most in your life—at university, perhaps—and think about how you learnt how to perform mathematical operations or scientific experiments through teaching and instruction—and not through the simple process of association. If either of these cases appeals to you then you are not alone. Psychologists have also proposed that there is more to learning than mere association. Indeed, we have already encountered alternative approaches to learning in the chapter on timing and spatial navigation. Cognitive psychologists have acknowledged that two systems might be responsible for human behaviour—two systems that the Nobel laureate psychologist, Daniel Kahneman, has referred to as 'thinking fast' and 'thinking slow'. The first system can be likened to associative learning. It is a system that does not involve any form of reasoning or deduction and allows organisms to, sometimes quickly, change their behaviour in a relatively automatic fashion. The second system is more deliberative and requires effortful mental processes, such as the formation of propositions about things and the use of logic and decision making. It may be that it is your subjective experiences of the second system that makes you want to acknowledge the presence of more than just mere association in learning—to want to explain behaviour, particularly in humans, in terms of more than one form of learning. However, as was noted in Chapter 1, subjective experience can be a hazardous tool for developing scientific principles. In this concluding chapter, I will describe some experimental studies that have been used to support the idea that there is a second, more deliberative, system in learning. This chapter therefore serves as a counterpoint to the idea that all there is to learning is association.

Follow the instructions

Verbal instruction can have a profound influence on human behaviour—simply being told that something is true (or false) can have a relatively long-lasting influence on behaviour and attitudes. This was demonstrated in a simple experiment reported by Stuart Cook and Robert Harris, who told human participants that whenever a green light would be turned on, then they would receive an electric shock. Cook and Harris measured the skin conductance (which provides a measure of anxiety) of their participants before and after the instruction, and found that the level of the skin conductance increased. Furthermore, instruction can interact with learning that is established more directly—by conditioning. Colgan, for example, gave participants a series of trials in which the illumination of a light was established as a signal for an electric shock—not surprisingly, this training resulted in the acquisition of a substantial skin conductance response. Subsequently, participants were given instructions that permitted them to be able to predict when the shock was not going to be given. As a consequence of this instruction, skin conductance responses declined. Together, these two experiments imply that conditioned responses can be both acquired and extinguished by instruction.

Some psychologists, such as Chris Mitchell, Jan De Houwer, and Peter Lovibond, have suggested that the effect of instruction on learning is difficult for associative theories to explain—in the instance of Cook and Harris's experiment, the light and the shock were never paired and should therefore not have had the opportunity to become associated. Instead, they argue that the effect of instructions supports the idea that participants engage in a process of propositional reasoning—that they deliberate that the light is a *cause* of the shock, rather than just something that is associated with it—and that deliberation and propositional reasoning are things that can take place following instruction or direct exposure to the stimuli. Personally, I have never found this argument too persuasive for two reasons. First, telling people that a light and a shock will co-occur may result in their representations being co-active; and telling people that a light will come on but that a shock will not, will result in representations of a light and 'no shock' being co-active. Under these two circumstances we might reasonably expect associations to form between light and shock and light and no-shock, respectively, and hence the strengthening or weakening of conditioned responding. Second, it is unclear (at least to me) why obeying an instruction necessarily has to result in propositional reasoning. For example, if I tell my daughter to 'Stop!' quite firmly at the side of the road, and she obeys my instruction, does it mean that she has engaged in some form of effortful deliberation? Or, has she simply responded to a stimulus? If there is any merit in either of these two objections then perhaps we need to look elsewhere for evidence of a more controlled, deliberative process during learning.

Perruchet's dissociation

One of the classic methods that psychologists use to determine whether a particular behaviour can be explained by two processes is to seek out a dissociation. For example, cognitive neuroscientists often look at the capabilities of patients with brain damage over a variety of different tasks. Sometimes they find that a patient has difficulty completing one type of task, such as a test of verbal fluency, but not another, such as verbal comprehension. This kind of difference—or dissociation—suggests that these two tasks are the consequence of two different processes—one of which is disrupted, the other not. Pierre Perruchet used a dissociation, but in a rather different way, to gather evidence for conditioning in humans also being the consequence of two different processes. In his experiment, Perruchet gave human participants trials in which a tone was intermittently followed by a puff of air to the eye. The important word in the last sentence is 'intermittently' because on some trials the tone was followed by the puff of air and on other trials it was not. Perruchet arranged his experiment in such a way that sometimes there were quite short runs of trials in which the tone was presented with the puff of air—for example, a run length of one would indicate a run of trials as follows:

Tone–, Tone+, Tone–.

On other occasions, however, there were quite long runs of trials with a puff of air; for example a run length of four would indicate a run of trials as follows:

Tone–, Tone+, Tone+, Tone+, Tone+, Tone–.

Perruchet was interested in how participants' behaviour changed depending on how many trials in a row the tone was presented with the puff of air. According to the associative account of learning, the longer the run of trials with a puff of air, the longer it is that conditioning is happening—so conditioned responding should strengthen as the run length increases. In contrast, we might expect something different to happen if people are engaging in deliberative processes. On a long run of trials with the puff of air people might hypothesize that because the puff of air has happened so regularly on recent trials then it is less likely to happen on the next trial. This is the gambler's fallacy—if something happens more frequently than normal over some period of time, then it will happen less frequently in the future. Consider tossing a coin—if you toss four heads in a row, you might be under the (false) impression that the next toss would be more likely to be a tails. So which of these results did Perruchet observe? In fact, he observed both. Figure 22 shows the results of his experiment.

Perruchet discovered that as the run of trials with an air puff increased in length from one to four, then the percentage of conditioned responses (blinks) increased from just under 50 per cent

Surely there is more to learning than that?

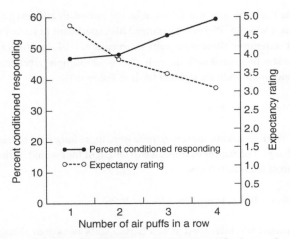

22. Perruchet's dissociation: the results from Perruchet's experiment. A dissociation between people's expectation of an air puff to the eye, and their conditioned responding.

to around 60 per cent, a result that is in line with the predictions of associative theories of conditioning. Interestingly, however, when asked to rate how much they expected the puff of air to occur following the tone, people's expectancies decreased as the run of trials increased. This latter result is more in line with a gambler's fallacy. This result is very interesting, because it implies that people's physiological responses to a conditioned stimulus can be dissociated from their consciously reported expectations. This is precisely the type of result that one would expect if learning was the consequence of not just one process, but two.

Even though the results of Perruchet's experiment suggest that two processes are required to explain how learned behaviour is acquired, one might question whether these have to be two, qualitatively different, types. Do his results require us to conclude that we have two systems of learning operating here, one of which is reflexive and based on associative principles, and the other of which is based more on deductive reasoning or propositions? Maybe not. Both results could be explained by one type of learning system. For example, perhaps conditioned responding is influenced only by the associative strength of the stimulus, whereas people's expectancies are more influenced by where in the sequence of trials the current stimulus resides. If we accept this kind of explanation for Perruchet's results then we are still agreeing that two different processes are required to explain his results, but we are not going so far as to say that one is based on associative learning and the other on something much more deliberative.

Logical reasoning in blocking?

One way in which psychologists have supported their claim that learned behaviour in humans (and animals) is a consequence of mere association is to show that the behaviour obeys the principles of an associative theory like the Rescorla-Wagner model, which is based upon the principle of association. For example, in an experiment conducted by Wolfram Schultz and his colleagues, it was shown how the learned responses of dopamine neurons in the midbrain of the macaque monkey could be blocked, just like the conditioned response to a light was in Kamin's experiment (see Chapter 3). This led to the conclusion that the behaviour of these neurons complies with the assumptions of associative theories of learning. But what if blocking was not the consequence of an associative process, but instead was a consequence of something more like deductive reasoning? If this was the case then such arguments by associative learning theorists would be undermined. This possibility has been proposed by a number of psychologists such as Tom Beckers and Jan De Houwer. Recall that in a blocking experiment, a conditioned stimulus is paired with an unconditioned stimulus (let's call this: A → US) in stage 1 of the experiment before the same conditioned stimulus is joined by another conditioned stimulus in stage 2 (let's call this: AX → US). At the end of the experiment, test trials with X typically reveal that rather little has been learned about X—it has been blocked by A. Instead of blocking being a consequence of associative learning, Beckers, De Houwer, and their colleagues have suggested that it is a consequence of the use of deductive logic. This logic or deduction is called *modus tollens*, and in the abstract language of logic it looks like this:

> *If p then q.*
> *Not q.*
> *Therefore, not p.*

Here is a more concrete example of the use of *modus tollens*, which might make it clearer:

> [*if p then q*] If I am an Olympic swimmer then I am very fit.
> [*not q*] I am not very fit.
> [*therefore, not p*] Therefore I am not an Olympic swimmer.

The same type of logical deduction can be applied to a blocking experiment:

> [*if p then q*] If potential causes A and X are both effective causes of a particular outcome, then the outcome should be stronger when both are present than when only one is present.
> [*not q*] The outcome is not stronger when A and X are both present than when only A is present.
> [*therefore, not p*] Thus, A and X are not both effective causes of the outcome.

The training in stage 1 of a blocking experiment clearly establishes that A is a cause of an outcome; it follows from the deductive reasoning above, then, that X is certainly not. Blocking should thus be observed through a process of deductive reasoning. Furthermore there is evidence that supports the deductive reasoning analysis of blocking. These experiments show that if human participants are given instructions or training prior to blocking that explicitly contradicts the 'if p then q' statement then blocking is undermined. For example, if people learn that when two stimuli are presented together then the outcome *is* of the same magnitude as when the two stimuli are presented alone, then the blocking effect is abolished.

It is, however, unlikely that blocking is always a consequence of higher order deductive reasoning such as the kind proposed by Beckers, De Houwer, and their collaborators. First, blocking can be observed in humans under circumstances when higher order processes are unlikely to be used, such as when people are given very little time to think about making a response on each trial. Second, blocking can also be entirely irrational. For example, if the number of unconditioned stimuli is decreased in the second stage of a blocking experiment from two unconditioned stimuli following A, down to one unconditioned stimulus following AX, then both animals and humans show an unblocking effect—all of a sudden they learn about X. This result seems difficult to explain with a system that requires organisms to engage in rational, deductive reasoning—why on Earth would such a system motivate an organism to learn more about a stimulus that is a predictor of less happening? This result can, however, be accommodated by associative explanations of learning that are not restricted by rationality or inference, and that instead are just based on the formation of links. In the case of the unblocking experiment just described, the omission of an expected US in the second stage of a blocking experiment is a surprising event. As you will recall from Chapter 3, surprise may not only enhance learning but also attention, consequently the boost in attention afforded by surprise allows learning to take place when it otherwise would not do so.

The question of how learning takes place is one that has concerned psychologists since some of the very first experiments were conducted in the field. The study of learning also has a tremendously broad scope, encompassing research conducted across the breadth of the animal kingdom as well as, now, artificial intelligence and machine learning. Despite this, as Nick Mackintosh stated in the opening line of *The Psychology of Animal Learning* in 1974, 'The study of learning in animals has a short history'. Consequently it should come as no surprise that opinions differ on what learning is and what psychological mechanisms underpin it. Relative to other scientific endeavours, such as physics, psychology is still an infant, so we should probably grant it an adolescent period of indecision and argumentation. In any case, the study of learning in both animals and humans continues to provide an insight into how behaviour changes as a consequence of experience. This insight is fascinating in its own right as well as having clinical and societal value. The study of learning should therefore continue to be at the heart of psychology for some time in the future.

Further reading and references

Introductory texts

Mark E. Bouton, *Learning and Behavior: A Contemporary Synthesis* (Sinauer, 2007). Provides an overview of many aspects of classical and instrumental conditioning.

John M. Pearce, *Animal Learning and Cognition: An Introduction* (Psychology Press, 2008). Introductory textbook on learning, broad in its coverage, including chapters on social learning, and communication and language.

More advanced texts

Anthony Dickinson, *Contemporary Animal Learning Theory* (Cambridge University Press, 1980). An influential book, providing a relatively cognitive analysis of classical and instrumental conditioning.

Mark Haselgrove and Lee Hogarth, *Clinical Applications of Learning Theory* (Psychology Press, 2012). An edited volume that focuses on how learning and conditioning can be applied to clinical conditions.

N. J. Mackintosh, *The Psychology of Animal Learning* (Academic Press, 1974). The definitive textbook on the study of learning in animals prior to 1974. Often the first place I look when I want to know the answer to something.

Chris J. Mitchell and Mike E. Le Pelley, *Attention and Associative Learning: From Brain to Behaviour* (Oxford University Press,

2010). An edited volume that investigates the relationship between learning and attention.

Robert A. Rescorla, *Pavlovian Second-order Conditioning* (LEA, 1980). A short monograph that manages to describe some complex experiments in a very clear manner. Includes some delightfully well designed experiments.

Chapter 1: What is learning?

Edward Lee Thorndike, *Animal Intelligence: An Experimental Study of the Associative Processes in Animals* (Macmillan, 1911).

Chapter 2: What is learned during learning?

C. Lloyd Morgan, *An Introduction to Comparative Psychology* (HardPress Publishing, [1894] 2012).

Chapter 5: When learning goes wrong

Eugen Bleuler, *Dementia Praecox, or the Group of Schizophrenias* (International Universities Press, 1950).

Chapter 6: Learning from others

Cecilia Heyes, Where do mirror neurons come from? *Neuroscience and Biobehavioral Reviews* 34(2010), 575–83.

Chapter 7: Surely there is more to learning than that?

Nicholas John Mackintosh, *The Psychology of Animal Learning* (Academic Press, 1974).

Index

Learning

ONLINE CATALOGUE
A Very Short Introduction

Our online catalogue is designed to make it easy to find your
ideal Very Short Introduction. View the entire collection by subject
area, watch author videos, read sample chapters, and download
reading guides.

http://fds.oup.com/www.oup.co.uk/general/vsi/index.html

SOCIAL MEDIA
Very Short Introduction

Join our community
www.oup.com/vsi

- Join us online at the official Very Short Introductions **Facebook** page.
- Access the thoughts and musings of our authors with our online **blog**.
- Sign up for our monthly **e-newsletter** to receive information on all new titles publishing that month.
- Browse the full range of Very Short Introductions online.
- Read **extracts** from the Introductions for free.
- Visit our library of **Reading Guides**. These guides, written by our expert authors will help you to question again, why you think what you think.
- If you are a teacher or lecturer you can order inspection copies quickly and simply via our website.